AAA* EINSTEIN...

AND BEYOND

By

Randall Barron

In search of the real

and understandable Universe

Including the Unauthorized Version

of Creation

And Our Ultimate Fate

*ACCESSIBLE... to the average person
*ATOMIC... covers the Quantum world
*ARTISTIC... comes together in a single vision

ISBN: 1-4033-4837-5 (Electronic)
ISBN: 1-4033-4838-3 (Softcover)
ISBN: 1-4033-4839-1 (Hardcover)

This book is printed on acid free paper.

1stBooks - rev. 07/29/02

PROLOG 1

Science today stands at a crossroads.

We have been brought to that crossroads in a strange manner.

Progress has not been at all in a straight line, or a spiral, or in any way that could be called smooth or orderly. On the contrary we are at times almost overcome by what seem to be conflicting views of what the Universe is and what its basic physical processes are.

Whatever the Universe is and whatever those actual physical processes are, they have not radically changed in the past few hundred years... only our perception of them.

Isaac Newton saw the universe one way, a way that many compared to the workings of a giant clock... something intricate and technically delicate, but basically understandable.

Albert Einstein pulled the rug out from under that comforting concept. Einstein was in fact a kind of bull in a china shop. Suddenly Time was flexible and empty space was somehow a construction, a curved and twisted thing almost impossible to visualize.

And then along came Quantum Physics. Just as people were beginning to try to adapt to Einstein's difficult vision. Now all horizons were drowned and there seemed to be no standards left at all, nothing to rely on. Even Einstein at last rebelled against the increasingly bizarre conclusions and indications of Quantum theory... stating in effect that "God does not play dice with the Universe."

And what are we to make of all this?

Do we the public just throw up our hands and basically despair of ever understanding the Universe in which we live?

Quantum Physics most advanced pioneers have often felt that way themselves. The quandary is a genuine and sincere one.

Recently... in the past 50 years... there have been attempts to simplify. Quark theory. String theory, and Superstring theory, for example.

But as yet for most of us living on this planet, the Universe and its workings remain a murky mystery, one whose resolving seems not to be anywhere on any visible horizon.

And yet...

And yet the mind longs for it.

And so the need for a kind of what I call *AAA Einstein*... Someone to step in and make sense of the situation in ordinary language for ordinary people. Not just to popularize others' conclusions, either, but to put in some of his own. Even, as it turned out, perhaps to penetrate some of Science's biggest outstanding and previously unsolved mysteries.

A real caped crusader...

IF genuine...

Because...

It would be a real comfort to be able to have a basic grasp of the kind of world we live in and the kind of Universe we inhabit.

Who could deny that?

I don't think we currently have that grasp. I don't think that kind of grasp is any way offered either by the bulky textbooks of Science, singly or in combination, or by any curriculum or teacher.

Question: Is it possible that a single missing element could unify all the disparate visions of what the Universe and its basic workings may be? And in so doing, make the Universe and its basic workings comprehensible to the average guy or girl in the street?

???????

Could actual pictures emerge from this unification, pictures which are not only simple to understand... but at the same time nearer to the reality that lies beneath all scientific theorizing?

That is the primary goal of this little book, along with a challenge to Established, Orthodox Science to take a look at what is being said here, too.

—Randall Barron

* * * * * * *

SOME BASIC QUESTIONS TAKEN UP IN THIS BOOK

1. What is Light?

2. What is Time?

3. What is Gravity?

4. How are the above three related?

5. Is Space curved?

6. How did our Universe originate?

7. Why is the scientific term "The Big Bang" a misleading one?

8. What fundamental questions did Albert Einstein leave unanswered?

9. What was Albert Einstein's biggest error?

10. Can a redneck amateur scientist possibly point the way towards solutions to Science's biggest unsolved mysteries?

11. Who was the first to propose that we live in an Expanding Universe? *If you state the name of any scientist, you are wrong.*

****

PROLOG 2

As a AAA Einstein, I feel obligated to give you even more of an idea of things considered in this book as a kind of guide. That way you can decide if you think it worth reading, or whether you want to skim through it quickly or ponder it bit by bit and at great length, or conceivably both.

Though the questions may be heavy, I have nothing against occasionally handling them in a light manner. It's my style as a carefree, Devil-may-care, shirtsleeved redneck AAA self-appointed scientist. True, I have been told by some readers of the manuscript that my style is cheap and flashy, even flippant and irreverent... and so, entirely inappropriate for a scientific book.

Well... maybe.

Or maybe my way of writing is just in reaction to having read myself too many deadly dull scientific books written in language altogether too opaque and abstract for my own poor powers of comprehension. And if I have a choice... well... I would always rather be too light and frothy and yet have behind that apparent froth some real substance to offer, than to write ponderously and pedantically just to please some mostly fictitious and certainly prejudiced idea of what scientific style should be... with nothing at last to offer behind the academic facade.

I freely admit I sometimes get what I might describe as creatively euphoric as I become swept along by the grandeur of the subject matter. Sometimes I later edit that out, but mostly no. I think spontaneity adds liveliness, a quality to be desired when walking through the sacred halls of Science, which all too often have been known to induce... slumber.

Plus, of course, *I am what I am*, and don't pretend to be anything else. Was that Popeye or God who was supposed to have said something like that? Well, I'll leave that particular investigation for the academics, which should provide plenty of material for one of their much beloved footnotes... of which my entire text has none. AAA Einstein is not much into footnotes, nor wishes to be.

Anyway, this prolog should give you an idea of the subjects taken up...

I cover one of Science's biggest current enigmas, which is this. *Why is the Expansion of our Expanding Universe being speeded up?* This goes against all science and common sense. Yet it has been confirmed by observation over a considerable period of time. So... what is going on here?

AAA Einstein gives you the answer, and in the process may have solved what leaves the best scientific minds of the modern world completely baffled. Judge for yourself.

Why do Gravity and Acceleration produce similar results? Einstein himself observed this and used it in arriving at his great theories. Yet he left something out... *the very important connecting link between the two.*

And, not such a small accomplishment, at least in his own estimation...

AAA Einstein tells you exactly what that missing link is.

And try this on for size...

Is it possible the Creation of our Universe still continues?

AAA Einstein gives you his no nonsense answer to the question that no one else even dares to ask.

Plus...

What is the true shape of Our Universe?

AAA Einstein tells you.

Examine this completely new concept...

What is a White Hole and why does it have significance for Our Universe?

AAA Einstein spells out the details in plain language.

What is the future of Our Universe? Is it destined to expand until it freezes all life? Or contract and eventually meet a fiery death?

AAA Einstein offers an intriguing alternative prospect.

What is the Governor and Regulator of the workings of Our Universe?

Read AAA Einstein's answer here.

What is the underlying truth behind the apparent constancy of the Speed of Light?

AAA Einstein delves into it.

Science points out that electrons inside the atom can take only certain designated orbits.

AAA Einstein points out the simple reason why that is true.

He also points out why Light can not consist of true particles. *If it did it would blow Einstein's theories sky-high.* Even though it was Einstein himself who insisted that light IS made up of small particles.

Then, consider this...

Does Our Universe contain a 5th Dimension?

Straight talk about this from AAA Einstein.

He tells you straight out that the constancy of Light speed is a myth and a misconception, or at best a trick of language. *That it is only the MEASUREMENT of Light speed that works out to be a constant in any local environment.* He tells you exactly why that is.

He also presents his idea that Albert Einstein's thought after achieving his first great accomplishments in Relativity Theory took him down a wrong road... a wrong road that to his own dismay only led farther away from his longed-for Unified Field Theory.

And, as they say, all this... and more... much more.

CONTENTS

PART 1

The Cosmos and Genesis. Plain and Simple from AAA Einstein...

Randall Barron

Saving the Best for First

I know I could save my conclusions for the end of the book, and so hope to engage my audience to such an extent that they would hang on to the last... being kept in suspense, as in some mystery novel.

But I do not think that to be appropriate here.

So I will take the reverse approach. First I will tell you how I visualize the Universe, in the briefest and most elementary terms. That simplicity is what my title proclaims and what I seek.

Then, afterwards, I supply what I think to be important details, answers to questions already raised... and points of historical background to all this as well as future speculations.

That way you get the major import first, and if your interest leads you to read on, well, then... we must be on the same wavelength.

So let's start there. What is the Universe like? What was its origin? What are the major forces governing its existence now? Everything answered in a simple, perhaps simplistic manner, in a few pages. Details... we can handle those later.

****

The Universe for Dummies...

First of all...

Include me in that title category. Please.

There is no doubt the term applies. The shoe fits, and so I wear it. Not with shame, but with pride.

I may be the world's first Redneck scientist. Or wannabe scientist...

Though if you pressed me about it, I might say at least two others preceded me, though their present fame tends to hide their own very unorthodox and uncredentialed beginnings. I speak of Michael Faraday and of Albert Einstein himself.

So...

If I make blunders, I only hope they are large enough to attract notice...

Here goes.

Science has come up with a theory of the origin of the Universe.

They call it the Big Bang Theory.

All I can say to that nomenclature is...

Shame on you, Science!

You dissed us all when you did that. Showed disrespect for the process, for the Originator of the process, and for us.

It is all right to be modern, to be informal. I myself am the epitome of that approach. But there is a limit.

I ask this...

Lose the term Big Bang.

I will not honor it here, except to express my own disrespect for it.

Besides being disrespectful, the term is also not descriptive.

Our origin is described as an explosion.

Think about that.

Does anyone have any positive associations with an explosion?

Think of New York. The twin towers of the World Trade Center...

I need say nothing more of such a ghastly tragedy, I know.

But... why then... why on Earth... should we ever adopt such a bizarre and derogatory picture as the basis of creation of our Universe?

I may be wrong, but to me an explosion is something destructive.

Not creative.

Is that too deep or complex for Orthodox Science?

My guess... yes.

And Big Bang? For any sound, you need air or something similar to transmit the sound waves. And anyway, as has been wisely asked... if a tree falls in the forest and there is no one around to hear, does it make a sound?

My suggestion is this.

Let us look at Creation in a new light. *Because darkness enough... we already have.*

Let me call on Einstein himself to rescue me here.

My contention being that the Creation of the Universe was not an explosion.

Imagine that.

Details will come later, but it is clear enough that Science describes the Origin of the Universe as coming from something like what might be described in modern terms as a Black Hole... but what I will call in this particular case The Primordial Mass... a density so intense as goes beyond anything we could easily understand.

But if we go back to that moment of origin, let me tell you, I CAN understand this...

About Time...

Time would literally stand still in such a situation. As modern science admits it does within a Black Hole.

So... if that is true, how could an explosion take place? An explosion being an instantaneous expansion. *Instantaneous?* In a Black Hole?

No way.

Not if you measure Time from within the confines of the Black Hole. Remember, that is where Time itself stands still.

Big Bang?

No way.

Creative Expansion? Okay. All right. Let's call it that.

Let's get our terms right. To avoid basic misconceptions.

That is my first point.

My second...

The PRIMORDIAL MASS itself from which our Universe sprang probably would not arise in a vacuum or a void.

It probably took place inside AN ALREADY ESTABLISHED UNIVERSE.

Is that concept too difficult for anyone?

If I use a tone of derision here, it is directed squarely at the scions of Science, and not at my readers. They and I are in the same boat the way I see it... a boat deliberately sabotaged before we got in with a thousand leaks from invisible holes.

Back to the Universe and its origin...

When its immense contraction was followed by an equally immense expansion, which may sound at first like an explosion, BUT...

Remember.

Said expansion DID NOT take place instantaneously.

Randall Barron

It took place through as yet unmeasured corridors of time.

In fact, it may still be going on.

Just remember you heard it here first.

Read my bytes. *It may still be going on.*

Which means... what?

It means we may still be *in the middle of The Creation.*

Did I lose anyone there?

Hope not.

Again I speak not of my readers but mainly of orthodox scientists whose minds in too many cases seem to have lost their normal elasticity...

It seems to me true. What I said.

We know the Universe is still expanding.

Science tells us that, and accurately. Yes, it may be only the remaining momentum of the original moments of the Creation, you say. But that is just my point. It cannot be. Why? Because the expansion rate now is ON THE INCREASE. Something that should not be happening, according to Science.

What is the explanation of that?

Well, for one thing...

It indicates clearly to me that the Creation DOES still continue. It isn't over. After Chapter 1 must logically come Chapter 2 in the continuing story. Even though we don't yet know what that Chapter 2 or any succeeding chapters might be or what they might concern...

So...

In that sense at least...

Perhaps we are, yes, still in the middle of The Creation.

But leave that for further discussion later on.

Let's introduce a question here.

What happens as the primordial material, The Primordial Mass, the Black Hole from another Universe, is propelled outwards from the central mass?

This is not difficult. A new Universe is in formation. What shape will it take?

Spherical. What other shape could it be? Dealing with radiation from a center that takes place equally in all directions...

And so we have the formation of our Universe described in terms so simple you can not find them... no matter how assiduously you look... in any other place.

So now we have established that. The basic shape of Our Universe.

A sphere.

Oh, yes, possibly slightly distorted due to the surrounding presence of the Original Universe inside which it formed. All right. No problem with that. No major one, anyway.

The sphere is, if we look around us... at our own globe, at raindrops, at the other planets, at the stars... obviously a favorite shape of Nature. There is no compelling reason I can see that Our own Universe should be any exception to the rule.

Stop and think about it, after all. No one contends the Universe is square... or a rectangle... or pyramidal... let's face it, corners of any kind are out...

If you are with me up to here, then so far, so good. Let us then move on to the next... pardon the expression... square...

****

The next square is not a square at all, of course.

It is not two dimensional... neither does it have corners.

No, it is a sphere.

That, to me, has to be the basic shape of Our Universe.

And if I have lost the attention of any auto mechanic who might be reading this book... then I have failed. Because what I want to do is make my theories and observations understandable... not necessarily to everyone, but at least to that auto mechanic.

Let us close it up here, then, as if 5 o'clock had come.

Tomorrow we can take up new challenges...

* * * * * * * *

Please note that during my next demonstration, at no time do my fingers ever leave my hands. Also, I do not break out into a series of formulas that Einstein himself would have trouble understanding.

No, this is your local AAA Einstein working here, with no pretensions towards being erudite, just trying to be plain and understandable.

But I feel I must make an announcement. I am about to solve a Great Mystery. One that has perplexed and plagued modern science for a number of years now. One that HAS no apparent solution.

This may not seem modest at all, or even permissible for just a self-professed and self-anointed AAA Einstein...

But...

I am going to do it anyway.

I am going to solve it for you on these ensuing pages. The solution will be simple, straightforward... subject to anyone's test of common sense.

So... as bold and possibly arrogant as it may seem...

Here it is...

Mystery of the Expanding Universe

There IS no mystery at first.

Everything seems clear. The Primordial Mass of immensely contracted matter of unimagined density... expands...

Matter is thrown outward from the Primordial Mass in all directions... a scenario that describes construction of a sphere, as previously pointed out.

The Progression of this Creation... because it is a Creation... goes like this. From the extreme density comes a phase of relatively rapid expansion... followed by slower expansion as the original impetus loses force, and as the Gravity of the expelled masses begins to work its attractive force, which is a brake on the continuing expansion. There is cooling, there is coalescence. There is the forming of stars and planets and galaxies. All right. None of this defies either common sense or the known laws of Physics.

But wait...

Modern science introduces THE MYSTERY.

Which is this. Light from the galaxies most distant from us presently indicates not only that the Expansion is continuing... we knew that... but also the astounding news that... it is ACCELERATING.

How can such things be?

They cannot. Not according to all the doctrines of modern Cosmology.

Scientists have looked for mysterious forces in the Universe that must be driving these galaxies to ever greater speeds. And have had trouble in finding any such mysterious forces.

But... what else could account for the acceleration?

Ah, yes... a question to be asked.

No one has yet come up with a viable answer.

But I have one.

I, the AAA Einstein.

And here it is...

* * * * * * *

I do love a mystery. To the extent that sometimes I hate to solve it, because to do so removes some of the original enchantment. But... I must push on. Sherlock Holmes and I have to put up

17

with our occupational hazard of suffering disillusion even at the height of the moment of triumph.

Ah, well. No going back now.

We were not created out of nothing. But out of something. That something was THE PRIMORDIAL MASS, an agglomeration of very dense matter. What we in our own Universe might compare to a super-condensed Black Hole. But though we can make and work with the comparison, that does not mean the two are equivalent. It is quite possible that THE PRIMORDIAL MASS was, from the beginning, more complex than might be any accidental Black Hole.

But, in the interest of comprehension, it does help to push on with the comparison.

Black Holes because of their extreme density, and consequent great gravitational attraction, suck in any matter that approaches them and do not let it again escape. In that way a Black Hole grows. Grows until... in some cases at least it reaches a CRITICAL MASS. At that point an explosive reaction can occur. But one, as I have pointed out, which takes place in slow motion.

And so something like that must have taken place with THE PRIMORDIAL MASS.

Which then expands outward radiating matter and energy in all directions... in the process forming a natural sphere which continues to expand outward.

Not a sphere standing alone in a void. Not at all.

It is a sphere within an existing context... within THE GREATER UNIVERSE.

Knowing that single fact changes everything, as you shall see, or perhaps have already leaped ahead and understood. As to explaining the mystery of the ACCELERATION of the expansion...

Now it all becomes crystal clear.

Once we understand that what we call and have called THE UNIVERSE, is really not that at all. What it is, is A UNIVERSE within another FAR GREATER UNIVERSE.

Now that revelation alone is well worth the price of the book... and had I any literary or economic sense I would have reserved it for the final chapter.

Because you can search a hundred books on Science and Cosmology... and never, ever find any such statement as mine.

Most scientists prefer not to ask the obvious question about the origin of Our Universe... about what existed in the moment PRIOR to their mistakenly called Big Bang. That is somehow considered to be... off limits.

But not for me. Not for your AAA Einstein, whose audacity by far outweighs both his judgment and discretion... and certainly any sense of political correctness, either within or without the world of Science.

Knowing that, let's take a new look at the so-called Expanding Universe which conventional science describes for us.

Yes, in its essence, on the simplest level, it is really something comparable to a Black Hole brought to its Critical Mass... one which has then expanded and continues to expand even as we watch and exist within it and try to understand it...

But now let's go to the next step. I have said we may be witnessing the continuing Creation of our Universe. And that, it seems to me, is true. *It is not over yet, far from it.*

So... what happens with the continuing, though slowing, expansion?

Slowing?

Obviously. Because with passing time, our Universe grows less and less dense. As far as we presently know, no new matter is being created. And what matter exists... increasingly occupies more space... and therefore is... less densely distributed within the forming sphere.

Remember, according to my script, we began as something like a Black Hole... within a much, much GREATER UNIVERSE.

Now... what happens when the lessening density of our own ever-expanding Universe becomes LESS THAN THE DENSITY of the surrounding GREATER UNIVERSE, the parent universe?

AAA Einstein has no hesitation in telling you what happens. None at all.

And it is information you will find *nowhere else...*

Modern Science does not DARE to postulate such a heretical, unorthodox thought. Not even Stephen Hawking...

What happens is the progressive conversion of what was once a kind of Black Hole... into what increasingly moves towards becoming what I would call... *a White Hole!*

Wait... what is that you said?

A White Hole?

????????

Well...

Don't let the terminology throw you. It's all very simple.

It goes like this...

A Black Hole attracts surrounding matter inexorably into it.

A White Hole begins to send matter inexorably out of it.

That is exactly what we see happening in modern times. That is exactly what has mystified modern science.

But the mystery becomes crystal clear with this concept.

We are a constantly expanding sphere living within a GREATER UNIVERSE, whose density now... because of our continuing expansion... is greater than ours.

This GREATER UNIVERSE surrounds us... it is everywhere on our perimeter. It should then come as no surprise what it does to us...

That is, it attracts all matter on our outer boundaries with an attraction greater than our own constantly weakening interior gravitational attraction.

What does that do?

Simple enough. You can figure it just as well as I can.

It ACCELERATES the galaxies on our outer rim... more correctly our outer shell. And in general it accelerates the Expansion of Our Universe.

That is logical, and can be clearly visualized and understood by anybody.

I visualized it. I understood it. I, the self-designated AAA Einstein.

And so, all egotism aside...

Am I too bold in saying...

Mystery solved...?

****

Moving forward in time, what will eventually happen?

Nothing very spectacular.

What I see is that our own Universe will gradually be integrated and absorbed into THE GREATER UNIVERSE...

A horrible destiny? As Orthodox Science is so fond of describing for us...

Quite the opposite.

Not in any way so bad a fate...

Since... unlike all the prognostications of Orthodox Science... my own does not include any foreseeable sensational disaster.

On the contrary...

Eventually everything will be brought to a level to agree with THE GREATER UNIVERSE, of which we will become an integral part.

So...

All right, there it is...

An explanation that any bright third grader could understand. Which gives a logical explanation for what has been puzzling Orthodox Science for Lo... these many years.

Thrown in with that solution of one of Science's greatest mysteries... is a complete thumbnail explanation of the origin and future fate of Our Universe. At no extra charge. All within a few pages at the front of the book.

AAA Einstein delivers.

And hopes you accept delivery.

These are things the original Einstein never remarked on, brilliant as he was...

Of course I will have more to say about all this later in the book.

****

PART 2

The Key to the Universe and Anything Ever Written About It...

Randall Barron

Onward

The preceding revelations... if that is what they are... in Part 1 are something I am most inordinately proud of...

I can hardly contain my pride within my own pitiful human context.

Due to the fact...

I believe it all to be true... and due to the knowledge that, as far as I know, no one has ever presented such a scenario before.

And add this...

To me at least, it all makes perfect sense and, as promised, does solve the mystery of *the accelerating expansion of Our Universe.*

I hope no one is too put off by my sometimes somewhat less than serious style. It doesn't mean the subjects are not serious stuff. They obviously are. Just that it pleases me to treat them lightly. It's an approach that lets me work right to the core of what I want to say... without recourse to scientific circumlocutions and excessive protocols and formalities.

I believe that even Science... or especially Science... should have room for a little self-expression.

Maybe that revolutionary point of view may be part of the contribution of this self-anointed AAA version of Albert Einstein.

Here I have to fire myself up again.

Because I am already suffering the Sherlock Holmes' sense of disillusion after a momentary triumph, and so I must move on to new challenges...

The next cogent question to ask is this...

What keeps it all together? The Universe... Our Universe, and THE GREATER UNIVERSE...

The answer is this... there is a unifying force.

This UNIFYING FORCE penetrates and permeates everything in the Universe.

It is invisible, so far undetectable. But it negotiates everything that happens or does not happen in Our Universe. It is the Master Controller, the Great Accountant in charge of all energy and mass exchanges. It is what connects everything and keeps it all functioning. The element without which we would not be here.

And while you could give it any name you might like...

Nineteenth Century Science had already figured some of it out for us.

They called it The Ether.

That is a term presently not in vogue. That is putting it mildly. Rather it is a term completely out of use. *Verboten*. Forbidden. Beyond the pale and beyond any possibility of government grant money. In fact, any modern scientist found guilty of using it would quickly be exiled to some Gulag of the scientific mind... even here in modern USA.

But I am not any modern scientist. I am AAA Einstein... your local redneck amateur mechanic kind of Einstein... And so I use it freely... The Ether.

That is what makes the Universe work.

Okay. Still, my definition of what is the Ether is not exactly what 19th Century Science said it was. I know that. And yet I want to give them the credit they deserve for basically having acknowledged it.

I have now done that, and am ready to move on to the next step.

I would change the name, would suggest THE UNIVERSAL MEDIUM to be the term of choice, as more appropriate... and, yes, also to cut off the undeserved opprobrium connected with the 19th Century nomenclature.

So... where do we go from here? What is next?

What is next is to describe our Universe... our world. And that is what I propose to do. In terms that may be different from those you have heard...

* * * * * * * *

The Universe is governed by the original projection of matter and energy from the Primordial Mass along with what was programmed within both.

What did I say? Programmed? But that implies, does it not, a Programmer?

Yes, it certainly does. How could it be any other way?

As to religious convictions, everyone can have his or her own.

But, scientifically... there must have been a Programmer.

That, I must insist upon.

The Ether... but let's call it now, as agreed, THE UNIVERSAL MEDIUM... was programmed to do many things...

To regulate, most of all. To maintain, and to be a kind of Stock Exchange of the Universe... at least an energy and mass exchange... maintaining within itself an ample supply of the basic building blocks of the Universe as well as the basic building waves of the Universe.

Once that is accepted, if it can be, the rest is basically easy.

I would see it as happening something like this.

As Mass differentiated itself out of the original projection...

Spheres formed.

Suns. Planets. Moons... asteroids. Comets. Galaxies.

There we have Our Universe. Simple enough to be understood by anyone. By my theoretical third grader even... or maybe especially so.

So next...

How would such a universe be regulated?

I have already said by The Universal Medium. All right. Easy enough to say. But how does that work?

The Universal Medium contains and transmits the force we call Gravity. How Isaac Newton and Albert Einstein interpreted that force is something I will take up at a later time.

For right now, it would help if you could accept this...

The UM is the basic overseeing force that controls everything else in the Universe.

Gravity is one of its constituents. It is carried by The Universal Medium.

Something else about Gravity...

The field of force we call Gravity tends to concentrate itself within and around Mass.

Any Mass. However large or however apparently inconsequential. If Mass is concerned, then so is Gravity.

But the UM is everywhere. Inside us, inside the earth, inside and around everything. It permeates all space.

It is in empty space and exists there in a certain standard distribution of lines of force.

But in a Mass... such as the earth... it is more concentrated.

That is the answer to one of the primordial mysteries of our Universe.

And if we can accept that, at least on a provisional basis, then we can move on to other important things.

* * * * * * *

So, I have already established that the Universal Medium is nothing less than the Keys to the Kingdom. It regulates and controls the entire operation of the Universe.

Details will be saved for later in the book. Right now I want to put it all together in the most simple way.

The Universal Medium's field of force tends to concentrate itself in and around Mass. That in effect is what Gravity is... a field of force concentrated in and projected from Mass and carried by the Universal Medium.

Where did the UM get this omnipresence? From the Primordial Mass from which Our Universe was formed. The UM was an integral part of it and was projected everywhere... resulting in what amounts to total permeation of everything.

Now let's talk about other basic things.

Time, for example.

What is it?

To consider that question, let's take two extremes...

In a Black Hole there is NO TIME.

Time is frozen, or practically so.

It stops, does not run, or hardly runs.

The opposite case would be when there is no matter at all anywhere. In such a case, there is no Time either. Any infinitesimal ghost of Time there might be would run so fast... a million years would zoom by in the space of a jet-propelled instant... so that in effect there would be no Time.

What is the common element here then in the creation and regulation of Time?

Mass. But not Mass itself so much as the field of force it concentrates within and around it. The field of force that is the Universal Medium and carries Gravity.

Does Time flow slower on the Sun than on earth? Yes.

The more Mass, the more Universal Medium concentration, and thus slower the March of Time.

What else affects the March of Time?

Acceleration. And speed.

Why is that? The answer is very simple. Acceleration builds up the concentration of the field of force of the Universal Medium within the accelerating Mass. To use a simple analogy, if you run with a kite against the wind it soars higher. Why? You are making more molecules of air strike the surface of the kite in a given period of time.

That doesn't strain anyone's brain, I hope. I speak here again more of the intellect of Orthodox Science than I do of any reader who like myself is of no already convinced point of view.

The accelerating Mass then accumulates Universal Medium lines of force within itself, in proportion to the rate of its acceleration. So that it acts and reacts as if it were more massive. And at really high speeds, approaching light speed, even if no longer accelerated, you have built up an amazing amount of virtual mass, enough to slow down and almost stop Time.

Back to Time again.

We have already seen that Time only exists due to the presence of Mass... which concentrates the field of force of the Universal Medium.

Acceleration on the grand scale, at one half the speed of light or more, slows down Time drastically... due to the extreme concentration of the field of force of the Universal Medium within the accelerating space ship or astral body.

You might want to visualize this...

If you are inside that space ship, two things at least have happened to you. Time has slowed to a walk, and you and the space ship have become more massive. Both due to the same cause... the accumulation and concentration of lines of force of the Universal Medium caused by your acceleration. You might now be as massive as the Sun as far as your effective or virtual mass goes... and if so, Time will march as slowly for you as it does on the Sun.

That to me does not now seem so hard to grasp.

And...

If you can obtain the Speed of Light, *you will have frozen Time.*

It... Time... will not seem to move, even though YOU move at incredible speed. And so, theoretically, you could circumnavigate Our Universe in close to zero time. And continue thus repetitiously Forever, if that makes any sense.

But maybe, even with all this explaining what CONTROLS Time... maybe I haven't yet gotten down to the specifics of *what Time itself is.*

If not, then here I try my best to simplify the concept... for you, but also definitely for me, who am not Einstein... but only some hopeful modern plebian simulacrum... a AAA version of the great

Maestro himself. And that only to date in my own perhaps ridiculously presumptuous opinion.

With that made specific, let me continue with my heresy or treason... or lack of comprehension... whatever it may be, but which I feel compelled to clarify as best I can...

Which is this, as best I can express it.

Time is a RATE OF VIBRATION. A frequency...

That may seem something of a contradiction in itself, since "rate" and "frequency" both imply a measurement by Time. Yes, it is all interrelated, which makes it difficult. But still, for me, that is what Time is... the pulse of things on the most atomic scale. Yes, it varies. Yes, it is controlled by Macro forces. But yes, too, at last it is what happens on the atomic and subatomic level that is Time itself.

If you can stop or slow that subatomic vibration, then you can stop or slow Time itself... we all beat with its rhythm and our lives are limited by it.

Of course I have not yet mentioned that one thing the Universal Medium carries and contains is an element of THE GREATER UNIVERSE, although in reduced and altered form. Something that we may call the 5th Dimension...

Which brings to mind a story of mine which makes the point in dramatic terms. So let's take a break here and precipitate what we have learned while we hopefully enjoy the story...

* * * * * * *

Duel Across the Milky Way

—Robert D'Artagnan

Twenty minutes to eternity.

Scott Diamond's right forefinger did a little hesitant dance over the chromed button that could save his life.

The enormous nebula in front of him loomed like a giant white spider against the blackness of space. A white spider poised to strike, because from somewhere out of its neon regions would come Jack Caesar's ship.

The vision of Jack Caesar's face, his one-time best friend, came to him. With a smile on its handsome features, which was not only customary but appropriate, because Jack was ignorant of the doom that awaited him. Of the conflagration the two ships would make when they collided near the speed of light in the black depths of space.

Served him right, Scott thought. To become the spider's prey in an instantaneous funeral pyre. Jack wasn't much better than a spider himself. To tell Jolene lies like that, behind his back. Which he was sure was the cause of her rushing into someone else's arms... the mysterious, long-haired bearded man at the launch site. Which if he hadn't seen it with his own eyes he never would have believed...

But... what did it matter anymore?

Maybe it didn't. Nothing else seemed to.

His mind went back in time.

Time, years... what did all that mean anymore? Time was a funhouse mirror, years could be ground up and spit out by the machinery of stars and speed until all that remained of them was meaningless space debris left to wander the incoherent universe.

Just that the mind... the mind grabbed onto old habits because it was the only anchor now in a sea of chaos.

Where? Where had it all gone wrong? How had the dream gotten twisted into darkest nightmare?

It must have had a beginning. Everything did, somewhere...

* * * * * * * *

The Army-Navy football game, 2020.

Late afternoon Autumn shadows beginning to dominate the green, chalk-striped field.

In the broadcast booth, the announcers were frenetic, ecstatic.

"Not since the historic duo of Doc Blanchard and Glenn Davis have we seen a backfield pair like this. Well, we saw those two only in old films, but these two guys..."

The color man picked up the gauntlet. "These two guys... while they're Navy, not Army like Blanchard and Davis, yes, are very reminiscent of them. But those two were Mr. Inside and Mr. Outside. These two..."

The play-by-play announcer came in again. "These two—the Shadow and the Flash—are everything for everybody in every dimension known to man and even to some housewives, if there are any of those left. Just kidding. But seriously folks... in the world of football backs, there just aren't any more like these two, and maybe never will be again... There's the snap..."

Down on the field, the blue and gold went into action.

"There he is, the Shadow, Scott Diamond, drifting along behind the line of scrimmage. Look at that, oh, baby!"

Three defenders had a shot at the halfback who seemed to be in no hurry as he looked for an opening. All three seemed somehow to have lost their bearings. One missed entirely, the second slithered off helplessly to one side, the third seemed to have scored a direct hit on the midsection of the Shadow, only to somehow end up on the ground while Scott Diamond continued his lazy promenade. Continued it, then cut inside to what seemed to be an uninhabited zone.

By the time it became inhabited, the Shadow was twenty yards downfield. Even when tackled, he picked up another five yards in the process.

The snap again.

This time the quarterback lateraled to the Flash. Jack Caesar took the lateral and dashed for the right sideline. Nothing doing there. The element of surprise definitely lacking. A cluster of Army defenders was there to meet him.

The doom of his run seemed written across their determined faces as they had him pinned against the sideline.

But wait. He stopped on the proverbial dime, spun around. As if he had afterjets, he blasted for the opposite sideline, leaving tacklers flailing at the air behind him.

He turned a corner, sliced inside. There was nowhere to go, but he went anyway. Then turned towards the sideline again.

Treaded and threaded it all the way to the end zone, tacklers flying at him all the way and connecting, but only with the air behind him...

* * * * * * *

47

Jolene was THE GIRL.

Scott Diamond, the Shadow, knew it the first time he met her.

Unfortunately that was in the company of Jack Caesar, the Flash, who simultaneously knew it too.

Not surprising, the situation, since Jolene was the daughter of a famous astronaut and already predisposed to favor someone like her father. Her immediate quandary became quickly defined to her. To choose between them. Desperately she looked for easy reasons to make the decision, couldn't find any.

Meanwhile, time went by.

She had no intention of playing one against the other, no matter how it looked. To them, to the media.

She loved them both. The Shadow for his dash and humor and human qualities and how he was like her father.

The Flash also...

* * * * * * * *

The Great Experiment was something that had no such name at first and came about by natural evolutionary processes.

An idea. A proposal. At last NASA put a practical base beneath it, with both a prognosis and a budget.

And the concept grew.

Two very special astronauts were always part of the project description from the very beginning.

The Shadow and the Flash seemed to have been created for the roles.

It didn't make Jolene's decision any easier.

She saw that, and so did they.

* * * * * * * *

It was Einstein who brought everything to a head. His theories and how they must affect the future of the human race. How this particular mission into deep space would put everything to the definitive test and let everyone know at last what the possibilities were. Of exploration, of life out there, of so many things...

No matter that three hearts had to be broken into pieces in the process.

But Scott Diamond was innocent of those consequences, at first. After all, he was an astronaut, not a theoretical physicist, and not at all an Einstein expert. He and Jack had won their space spurs mostly by being the best of the group in flying the Moon Shuttle. They were throwbacks to the old breed of astronaut, pilots first and always and anything else a bonus but not of the essence in their own minds.

Scott remembered the day the first doubts of what they were going to attempt to do entered his mind. Entered it and lodged there with the immense stark clarity of Arctic ice impacting that famous historical transoceanic liner, that what's-its-name... Atlantis... no, wait, Titanic...

He should have known in that minute his life was headed for disaster. Should have backed out of the assignment, let someone else do it. Would have, but then his whole life had been built on the idea of plunging ahead, of overcoming difficulties like he used to overcome tacklers. By heading straight at them, then at the last second becoming the Shadow, the massless dancer who could somehow improvise a maneuver that would cloud the adversary's mind and avoid all harm...

He should have known those tactics wouldn't work. Not when the adversary was Einstein, and what Einstein saw in space travel that he did not...

****

"Only three years will go by for you on the trip out. But because we're going to get you up very close to C... light speed... earth time will be... different."

The blue eyes of Anderson, the Einstein briefer, seemed to Scott Diamond to shift sideways so he could not make contact with his gaze. "Different? What do you mean, different?"

"Well, it's a key part of the experiment, really its raison d'etre."

Already Scott wasn't crazy about Anderson, and the French didn't help. He thought he noted a similar reaction in Jack Caesar, but wasn't sure.

After an awkward pause, Anderson went on. "I thought this had already been explained to you...?"

"No," Scott said curtly. They had been told by NASA it was a six year mission. Three years out, three years back.

"Well, if Einstein is right, then... while your time is greatly slowed, time on earth will not be and consequently..." He raised his hands as if the answer were obvious and did not need stating.

Scott's voice was hard, relentless. "Consequently what?"

"Well, no one knows for sure. That's the point of the experiment. And it depends."

Jack Caesar got his two cents worth in. "How <u>much</u> time will elapse on earth?"

"As I said, it depends. On your speed... on whether the Einstein equations hold in deep space. But... could be fifty, five hundred, even five thousand years. It's all a question of mathematics..."

No, it wasn't, Scott remembered thinking at the time. Not at all. No, sir.

To him, then, and to him, now, speeding at near 180,000 miles per second towards the enormous phantasm of the glowing nebula, it was a question of something else. Of something very different, indeed. Of treachery and deceit, of misrepresentation and fraud, and of deliberately downplaying possible negative consequences.

The resentment had come out in his voice. "We were told... Jack and I, it would be a six year mission. Three years out, three years back. Simple as that."

Simple and heart-rending. Those were six years one of them, whichever she chose, could otherwise have with Jolene. Loving her, starting a family, living a normal life, creating a future...

Anderson was in trouble, but made an obvious effort to right the sinking ship. "Three years one way <u>astronaut</u> time. Yes, that's absolutely correct. What the earth time will be... nobody knows for sure... maybe just three years. In that case, you'll both be back here in six. Exactly as you were told."

Scott was going to say something, he was not sure what, but Jack leaned forward, beat him to it. "But...?"

"Well, that's where the theoretical considerations come in. See, if we can really get you up near the speed of light, time would almost stand still for you. In fact, if you got to full light speed... which, by the way, we <u>don't</u> think is possible... well... you'd be immortal for as long as you maintained that speed. You could circumnavigate the Universe in zero time. Zero time for you, though maybe eons for earthlings, depending on how big the Universe really is. And its shape, because maybe circumnavigate is a completely mistaken term..."

An idea like a butterfly in spring sunlight skittered across Scott's mind. "And what if... what if you could go faster than the speed of light?"

Anderson smiled, but Scott thought it was a very uncomfortable smile. "Why not ask me what happens when an irresistible force meets an immovable object? How many angels can dance on the head of a pin? Or what is the sound of one hand clapping?"

Scott was not about to give up so easily. "Yeah, but..."

"Okay. It's supposed to be impossible. Infinite resistance, infinite mass... that's what the space ship would build up close to the speed of light. So, zero chance. No chance at all of going past the

Randall Barron

speed of light. That's what Einstein says, that's what we say."

****

After that traumatic day, Scott maybe would have given up the mission to someone else. Would have except for two things. He was constitutionally averse to giving up anything challenging, plus there was the thing about Jolene.

She was after all, an astronaut's daughter. She believed in the mission, believed in what seemed almost to be the forces of destiny.

Because she refused to choose between them.

That decision she was ready to defer until their return from deep space.

The media made a big thing of that. Made a big thing of the fact she was willing to wait seven years. Most media pundits thought that was generous, that six years would see the astronauts return. Although they professed to believe

56

Einstein's theories, they drew a line when the theories applied to actual human beings, media darlings. They simply could not lend credence to the idea their creations could outlive them a thousand years, regardless of how some people might interpret Einstein.

Meanwhile, the proposed Great Experiment grew more bizarre.

The two space ships would approach each other from opposite vantage points in space some two hundred millions of miles apart, each at near light speed.

That would put all the most extreme Einstein postulations to the acid test.

What would happen under those conditions?

What would the light-speed measuring apparatuses on board each ship record as to the speed of the other ship?

Would they show, as common sense dictated, a relative speed of C times two? Twice the speed of light? Many reputable scientists said yes, because recently there had been good reason to call Einstein's favorite hypotheses into question on the subatomic level.

Though Einstein himself would have said the relative speed between the two ships would be recorded by each ship as C, exactly equal to the normal speed of light, regardless of their apparent doubling of that result.

And piled on top of that quandary...

Each ship was to have a laser light to slice through the blackness of space. So what would be that result?

That is... two ships headed at near light speed towards each other would seem to test the Einstein formulas about the limit of velocities in the Universe to the logical extremes.

Because... now you add to the basic situation the fact the laser lights of each ship are also traveling towards each other. Not just at light speeds. No. Because if you add the velocity of the space ship to the speed of the laser light on each ship...

What do you get then? If each ship measures not only the speed of the approaching other, but also is equipped to measure the speed of the laser light from the approaching other...

Whatever results you got they would be not just earthshaking, but reverberations both universal and revolutionary, whatever the measured outcome...

* * * * * * * *

The day of the launch came.

The inevitable day. August 28 of the year 2028.

Scott wanted to treat it in his mind as any other day.

But how could he? When the world could not...

When every atom in his own body could not.

And if you wanted an analysis of when things went finally wrong, all you had to do was X-ray that day.

That would tell you why Scott now still had his nervous right forefinger doing its dance over the chromed button. The electronic rudder override that still, in the twelve minutes left to him, could save his life...

But the analysis gave him no reason to change his decision. No reason now to make him press the button and avoid the final collision and conflagration. The ultimate disintegration.

Because it was all Jack Caesar's fault. There was no other explanation.

Jolene had been resolute. Beyond change. And then before his own protesting eyes, Scott had unwillingly seen her change. Why?

The only person on earth besides himself powerful enough to cause that was Jack Caesar.

That was why he hated Jack Caesar. Why he had to end his life now, even if it meant ending his own and causing the end of what had been conceived as a noble experiment, the Great Experiment.

* * * * * * *

It came back to him now in bitter dregs he felt deep in his blood.

Launch day from Moon Base.

The nurse had come into his office to inoculate him. Just to prepare him for the second, deeper inoculation which would put him to sleep for two years and eleven months. Because NASA believed that while the space ship endeavored to build up velocity close to that of light, the two astronauts could best survive by being prone and asleep in their webbed and molded cocoons.

That was preparation for what was ahead of them in space.

But nothing could have prepared Scott for what he saw out his office window shortly after the nurse left.

In the opposing second level window, the observation window of the Space Bar, the semiprivate hangout of the astronauts and all the NASA people...

It was Jolene, no mistake about that. That was all right, because he knew she wanted to be present for the launch.

But as she watched, someone came up beside her.

The long-haired bearded man.

That would have been all right except...

Except he immediately engaged her in what must have been riveting conversation. Because she obviously gave him all her attention.

Was it some hallucinatory effect of the inoculation, or did Scott really see what he thought he saw next?

He had no doubts about it now.

The bearded stranger put his arms around Jolene's waist as Scott watched, bent her backwards, kissed her...

With a proprietary air that told Scott one thing only.

His own life, his goals, his dreams, were moonbeams... no more substantial than that... his personal future cancelled out by the picture he saw before him.

And as to how that could come about, he had little doubt. Somehow, some way, only one person could be responsible for it.

Jack Caesar.

But there was a remedy for all things.

In this case, Jack Caesar would suffer the entirely justified consequence. What kind of lies had he told Jolene to cause her to betray them both?

Or had it all been staged just for his benefit, just to break his heart when it was too late to do anything about it?

How Caesar must hate him, the man he had once been proud to call friend... and how long had it been going on behind his back, this Janus-faced villain plotting and scheming against him?

Probably longer than he imagined.

Well, nobody did that to Scott Diamond.

Not without consequences.

And so, Jack Caesar would be obliterated from the screen of human existence.

There might be more to it than that, some deep secret explaining why Jack would suddenly turn traitor. Not that it mattered in the moment to Scott Diamond. All he was concerned with was his former friend, the Flash, would be eliminated by the Shadow, in one final monumental stroke of poetic justice.

And if the Shadow himself had to die in the process, what was that? If there was no longer any Jolene to validate him, to make him believe he was who he was, then life had no meaning anyway...

So might as well go out in a blaze of... a blaze of what? Not glory. No, call it a blaze of vengeance.

He decided to enjoy his last moments, savor them down to the last irrevocable final microsecond.

He walked over to the small refrigerator, twisted the cap off a Rolling Rock beer. No preservatives, the little sticker on the green bottle reminded him, and he allowed himself a bitter grin and a low chuckle.

He sat down on the large upholstered sofa positioned some thirty feet from the front observation window and took a healthy swig of the beer.

A blue light blinked and all three monitors announced another space ship was trying to contact him.

Forget you, Scott thought, and turned the temporary override to silence the blinking blue light.

He let the bizarre scene and situation sink in fully, now that he had no heed for being successful in anything but complete and utter destruction of himself and Jack Caesar.

In a way he was King of the Universe. Travelling in a big Luxury Liner of the Stars. But why not, the engineers at NASA had told him. Blasting off from the Moon Base, there was no reason to try to cut down on weight or size of the ship.

In fact, the ion engine designed to give him perpetual locomotion needed a wide base to scoop up the flotsam and jetsam of space, and if he walked forward and flattened his nose against the three foot thick maxiglass of the forward observation window, he could just make out the enormous shark's jaws below him on the front of the ship waiting to devour anything that might come their way...

He took another swig of the beer, found it incredibly refreshing. He was glad he had opted for the Victorian decor of the large combination sitting and control room. It was homey, reassuring.

Just what he needed before dying and dealing death to a traitor. Well, what was death? Nobody knew, but everybody said it was like a journey of exploration. If that wasn't a piece of cake for an astronaut, what was?

Eight minutes to eternity...

The stars reeled by on either side, and the giant spider of the nebula ahead loomed ever larger, now filling the field of vision through the observation window.

Off to the left and out of one of the whorls of the nebula, there suddenly appeared a slim red fiber reaching for his space ship.

Of course, the laser, Jack Caesar's laser. Right on time.

He watched the results come up on the big secondary monitor screen at eye level on the right hand molded wall of the ship.

Well, Einstein, wherever he was, must be pleased. That laser beam travelling at the speed of light and coming from a space ship also travelling very close to the speed of light and meeting his own space ship also travelling close to the speed of light... all three velocities added together should come out to three times the velocity of light.

But the monitor screens showed clearly they did not. The laser clocked in at an even C, no more, no less.

Too bad the boys at NASA would never know the results. Or maybe they would in another three years. They didn't need his, Scott Diamond's return to tell them, since the signals of the result would be automatically bounced back towards earth.

The big monitor screen was announcing something else.

MAIL CALL it said.

Good Lord, so they had done it. Maybe.

It was a surprise, because it had sounded crackpot when the NASA boys explained it, or tried to. To hang a laser God and Einstein and the NASA engineers only knew how many light years out in space and then make it stay there circulating in a confined area... it boggled Scott's mind.

Now he was going to intersect with that confined area and pick up all the news from earth. Like a train scooping up a mail packet off a station hook. If it worked.

Scott didn't care whether it did or didn't. And then when the monitor signaled incoming mail he suddenly found he cared very desperately.

Jolene. Maybe... maybe there would be something about her...

There was.

It flashed by quickly, as all the news did, as if in fast motion forward and the computer did not respond to his attempt to slow it down.

But he caught the essence of it. Jolene had married, less than a year after their blastoff. She was Mrs. Woodash now. The husband of the long-haired, bearded guy who had given her the passionate kiss he had witnessed at the moon base just before lift-off. The news photos of the wedding flashed by fast, but the face and the name of the scavenger who had made off with Jolene burned into his brain.

What had Jack Caesar done to cause that?

Scott thought he might have it figured now. Jack, despairing of winning her, had acted to make sure he, Scott Diamond, could not win Jolene either. Had told her some colossal lie that had turned her against him forever, made her decide a plague on both your houses...

Well, Jack would pay for that. Soon. Very soon. In... a little over three minutes.

Now the news on the monitor was really going crazy. Slashing by so fast he could hardly make any sense of it. What was happening?

He remembered the controversy about it. Some said the news channel broadcast into space and out to the distant astronauts would be seen as running slower than normal, others had said it would be faster, especially as the astronauts travelled further and in the process neared light speed.

Well, the answer was in and the speed boys had won.

And he saw something else. The time elapsed on earth, as the news flashes—and that name was eerily apt now—showed. It wasn't three years. It was thirty as they kept flicking by, the reports. He made a mental note of the terrible World Stock Market crash of 2037.

What did he care now that Jolene was lost to him, now that he and Jack had... what? Two minutes to live? He checked to see he was still locked into collision course.

And then...

My God! It could not be true. Something so terrible... something dreamed in the depths of night for so long but never... never had Scott ever thought it could really happen.

There were flashes about the Atomic Bandit Groups and then...

Suddenly there were camera shots from the Moon Base, horrifying shots, of the earth. But not the earth he had known.

No, this was a brown disaster, misshapen and offcenter and wobbling erratically...

Scott finished the beer in a gulp, felt something electric go all through his body. No, it could not be true. But... was.

Good Lord. Lord of hosts, Lord of... but no, no Good Lord could permit... that...

The monitor screens went blank.

What did it matter? One minute to eternity for him, too.

And then it hit him. Hit him through the deep shock his body had fallen into.

He was a fool to kill himself now. To kill Jack Caesar.

He suddenly remembered what the NASA shrink had told them. Paranoia. Normal maybe to feel it after the long sleep and the lonely awakening in deep space. Maybe Jack wasn't responsible. Not responsible at all for what happened to Jolene...

And anyway they might be the only earthlings left alive except for the Moon Base people, and maybe even they with the earth out of whack...

Jack, his buddy, his fellow backfield immortal, his friend and drinking companion. The Flash.

Thirty seconds to impact!

Scott turned slightly to his left on the sofa, to the ship's electronic control panel. Let his right forefinger fall against the chromed button, which inactivated the collision course order and automatically punched in ten degrees right electronic rudder...

* * * * * * *

Light is the media that permits sight. Scott Diamond knew that well enough.

But when something approaches you at the speed of light, the velocity itself works against your actually seeing it.

Scott felt it though. Jack's ship.

So close the shock waves of the near collision shook his own ship like an instantaneous earthquake.

Nothing to see except a blue flash. And it was all over.

That was the thing, though, Scott realized. It wasn't over. It was just starting. Because somebody had to. Had to make a new beginning. Something loomed in the dark recesses of his reeling mind. An idea, a plan, except without shape or form or anything like words...

He grabbed the communicator, spoke into it... "Jack... Jack... talk to me, talk to me."

He didn't know what to expect. Einstein again and his critics. Some said two space ships separating at light speed could not make electronic contact...

Jack's voice came through as ragged and broken as anything Scott had ever heard. But he heard it like a requiem. No, not a requiem, either, more like an Easter morning sermon.

"Scott... did you see the news... terrible... can we do?"

"Jack, do you remember Anderson back at NASA?"

"Anderson... yes... what he said... Magellan... maybe I could try for that."

Scott understood in spite of the static and the lapses. They were on the same psychological wavelength after all. Just as they had been so many times back on the football field. Almost as if no time had passed since then.

The voice came through again, weaker and even more broken up. "Close to the speed of light... zero time... circumnavigate it... immortal... I could go on forever, pal. The idea..."

Intrigues you. Yes, I understand, thought Scott. Forever on the cusp of the moment, forever young, forever slashing around the universe like he used to slash around the football field, elusive and always out in front... it fits.

And me? What about me? Well... if Jack can hear me...

"I read you, Jack. Perfectly. My idea goes back to Anderson, too. What if I could break through? The light barrier, I mean. Go faster than light."

Static and then. "... can't be broken."

"They said that about the sound barrier, remember? I know I'm talking Einstein here, against him, but anybody can be wrong, even Albert. Anyway... what have I got to lose?"

For a long minute there was nothing. Then, incredibly weak and broken, but understandable..."Good luck. So long, Shadow."

There seemed to be something in his throat, but Scott overcame it to finally push out the words and publish them in space.

"So long, Flash."

* * * * * * *

The name he hadn't mentioned was Jolene. Yet Scott Diamond knew she was the nexus of his whole idea, in a way he still hadn't gotten a theoretical handle on.

He wondered if somehow Jack's idea in some way he didn't understand related to her, too.

Well, time would tell.

Ha. Time was a joke. An Einsteinian joke, a Quantum joke, whatever... Whatever it was or used to be was beyond his powers to comprehend or work with now.

Yet that was what he had pledged himself to do. Work with it. Somehow. Some way.

In the hopes of...

No. If you spelled out the hope too clearly, too explicitly, then, by some unknown law, that increased the probability you would never attain to it.

Was that his own idea, or was it something he had stolen from the ancient author, Hemingway? Or was it much more probably just rank superstition?

He would not trouble his mind with it anymore. Instead, he would just do it, and if the idea prospered, fine. If not... he couldn't lose any more than he already had.

At least that was the way he chose to look at it.

* * * * * * *

In Scott's dream Jolene's breath was warm behind his right ear and his arms were clutching her tight, tight while he breathed in the perfume of her dark hair, let it intoxicate him. The dream seemed frozen there, or rather suspended in a kind of eternal cloud of joy. In the situation he did not question that... it was the kind of eternity he wanted... the only kind.

But then the shaking began. He woke to it, knew what was happening, or thought he knew. Back when the sound barrier was first on the point of being broken, that was typically what happened. The ship vibrating, shaking as if it were to come apart at any second as the sound barrier was approached.

Now it was happening to him. In spades.

It was days since he had said his last goodbye to the Flash, days since he had done what might prove to be his last rash act.

He didn't care. This would be it, heaven or hell, make or break or maybe nothing but death or doom. Okay. He had done what he had done and he wasn't going to back down now.

It was easy enough to do. Open the side wall panel and start to work on the little mechanical restraints that acted as a governor on the throttle mechanism. That let him pull the throttle all the way out two inches beyond the MAX mark.

And though it took days for the final burst of speed to build up, at last it had, and here he was, pounding at the invisible, massive walls of the lightspeed barrier.

He twisted the cap off a beer, then let himself be automatically strapped inside the transparent egglike cocoon and looked out at the stars, misty uncertain presences. The great nebula was somewhere behind him now, left in the wake of his light-like speed.

Science fiction writers and a few scientists had babbled about hyperspace and wormholes and who knew what else. Scott himself had never taken much stock in what any of them said. Instead, in the last weeks before launch he had applied himself towards working out his own theories.

What he came up with was something uniquely his own. He didn't know if it was deep, or just plain dumb, but at least it was his and made sense to him.

He went all the way back to the nineteenth century for his inspiration. The old concept of the Ether...

Maybe that idea was not so dumb after all.

Einstein had never denied its existence. Scott figured all it needed to make it work was the concept of elasticity. There might really be an Ether Web or Cloud penetrating everything in the Universe, holding things together, acting as a kind of energy exchange and maybe even a kind of Central Processing Unit.

If it were almost massless and flexible, then most of the Einsteinian concepts of warped space could still be explained, just in a different way. That is, any mass could hog a big share of the local Ether cloud, if mass were a natural conductor and concentrator of it, just like iron could concentrate lines of magnetic force.

Speed too, would force more of the elastic web or cloud inside the moving object, making it in effect more massive, just as Einstein said. He worked out the whole thing, and to his own mind, at least, it computed. Besides, he could practically visualize it and it just seemed right.

As to Hyperspace or whatever you might want to call it, his instinct told him it might really be there, somewhere. And he figured that somewhere most probably was on the other side. The other side of the light speed barrier.

Sure, maybe Einstein was right. You couldn't go faster than light. Not in the normal confines of space and time in the visible universe.

But that was the whole point to Scott's theory. Once you broke through the light barrier you no longer would be in the normal confines of space and time in the visible universe. You would be outside them, and maybe at the same time, depending on your point of view, on the inside of the inner workings, the master controls and mechanisms that kept everything in proper operating order.

The price for being wrong about that was nothing much. Just his life, he thought. Just total annihilation.

He nipped again at the bottle of beer.

Even inside the cocoon where his re-inforced and shielded body was held in webbed suspension delicately controlled by a gyroscopic mechanism... even in that safe harbor he felt the shaking growing worse, the ship surely on the point of coming apart now.

Whatever would happen must happen soon, because the physical construction of the ship simply could not resist much more of this...

The other side. In some way, in some sense, he was going to find out something, and very soon...

BAM! BAM!... BAAARAAAMMMM!

The shaking became catatonic. An infuriated giant had the fragile spaceship in his hand, shook it like a castanet...

Scott covered his ears from the crescendo of sound, his eyes from the unknowable...

Then...

Stability. Smoothness.

Silence.

Of the tomb. Was that it? Death in whatever form?

Scott feared to know the truth.

Opened his eyes anyway.

To what outside the ship seen through the viewport seemed now something like a waterfall of yellow light.

He knew he was somewhere else.

No stars.

He didn't know where the light was coming from. Nothing he could see. He felt more than observed he was in a corridor of some sort. A gigantic corridor whose curved walls he seemed to feel were out there. It was like a cave of space, an enormous cave or chasm he had broken into. The yellow light seemed somehow familiar, something he had maybe seen once in a dream.

He released himself from the cocoon, staggered to the sofa, peered ahead.

Suddenly he saw the walls. Translucent, almost transparent, but distinguishable. Dead ahead. He swerved the ship away from them.

As he came out of the turn, he became conscious of the clocks. They were doing strange things. The ship's clocks were stopped dead, while the earth clock began turning counterclockwise.

"Going backwards," he breathed to himself. Even as he watched, the backwards spin became more pronounced. "Attaboy," he heard himself say. "That... that's the way. The way home."

His lips had said what he hardly dared to hope.

But it was time to do so. Sure. Sure, that was what he was after. Home. Home and Jolene. Nothing less.

He called up his star maps on the computer screen. Quickly he tapped in the black button labelled "earth", then left the ship to Automatic Pilot.

He looked at his speed indicator. Still slightly more than C. Okay. Earth's clock was still spinning back, while the ship's clocks were stopped. He had to do some coordinating now between time and space. He set that into the computer as a special order which he constructed now by hand. DESTINATION, AUGUST 28 OF 2028, PLANET EARTH."

The computer would do the rest, even if it was a new and different task.

That was his hope, though he pretended to himself there was no doubt about it. As if inventing a time machine, jury rigging one, were an everyday occurrence not to be marvelled at.

He found himself whistling tunelessly, looking obliquely at the star map on the computer screen with the spaceship's position and the earth clock as if unconcerned about them...

Sucked in his breath and found the whistle had become a recognizable tune as he saw things moving the right way.

He had a chance. A chance of making it really happen.

Maybe things could be saved. Called back. Changed.

Jolene's rendezvous with the long-haired, bearded stranger. Maybe he could be there, intercept the stranger, convince her of his love, make her understand...

Fight for her. Win her back. Survive, live, endure. Live out what should have been his destiny with her. Make that destiny real again.

What was this place? This spinning continuum he had blundered into on a sword of speed? Was it what the scientists on earth had guessed and gew-gawed and hemmed and hawed in print so much about?

Was it really Hyperspace?

Or a Wormhole?

Or something else...

Some kind of interstice in space and time where all kinds of arcane business was carried on? By who or what? The blind organizing force behind the universe? God? Or at least the inner workings of part of the mind of God, as the twentieth century scientist Stephen Hawking used to say...

But now it was time for he himself to put up or shut up.

Be or not be.

Go or be cancelled out.

Become an integer or be an eternal zero.

And zero was a number he had no affinity for.

It was a marker, a place occupier that was against the nature of the Shadow, unless it were just a temporary disguise, a momentary deception with which to ward off tacklers.

One thing for sure. He didn't belong in a wormhole since he was not a worm. On the other hand, if it was just a passageway to paradise, count him in... please.

The monitors announced it. The coordinates had come together. Time for him to act now.

Now... or never.

He was poised and ready to dive for the cocoon again. But he had it figured without knowing why that the exit from the corridor would not generate the enormous shock waves the entry had. This time he would not be breaking the light barrier, but coming back from its other side.

He hoped.

He shoved the electronic rudder hard over. Saw the translucent walls ahead rushing at him headlong.

Damn the torpedoes!

Then sparks... a million sparks... inside, outside... even in his brain.

And...

Breakout!

The presence of the stars announced it, the... the moon! Earth's moon visible in the near distance!

He turned on the rear view screen to see if he would see the mysterious enclosure he had just burst out of.

Nothing. No trace of it.

He wondered how it could be. Such a vast chasm when you were inside it... and now... where was it?

No time to think about it.

He saw the grateful ball of earth below him. Not brown and wobbling, no, but like it used to be. Blue and whitely capped by clouds and inviting like nothing else in the entire universe.

And the earth clock showed him what he longed to see. August of 2028. Early on the 28th day.

So there was a chance...

But only if he could get his velocity down to almost nothing so he could land on the moon.

To do that, in the limited time span he had, he would have to perform a desperate maneuver...

He saw the speed indicator was now on the safe side of C. Sure, it had to take a load of energy to break out of whatever it was he had broken out of. But the earth was rapidly sliding out of his view...

Quickly he described to the computer genie the maneuver he wanted to make concerning the sun, keyed in all the parameters, then shut down the main engines.

* * * * * * * *

What Scott hoped he would be able to pull off was this...

Change course to intersect the sun itself. Of course he wouldn't, not if he were lucky enough for the plan to work, and if the ship's systems did not fail him.

In a way it would be like reliving his football days. Back then he perhaps more than any other player had been a master of evasive tactics. Certainly he always had an intuitive feeling for when and how to exercise those tactics.

Like now, maybe.

Only the tactic he wanted to exercise was more Jack Caesar's than his own. Jack was the master of it. It was the Flash more than the Shadow who had made it his particular trademark.

To be able to turn on a dime when confronting what looked like an impossible situation and...

Reverse field.

And precisely that was what Scott would try to accomplish now...

That is... approach the sun as close as he dared, then treat that august body no better than a sideline or a charging linebacker would be treated by Jack Caesar.

Surely the Shadow could still do that, no matter that the playing field were as enormous as the stakes...

Scott knew it was his one chance.

By reversing his field then the sun would instantly become the brake he so desperately needed. The sun's enormous decelerating power as he turned away from it and back towards earth would be the key to everything now.

To the possibility of his eventual safe landing on the moon, and then...

He felt his racing heart skip a beat... and then... Jolene. His future. Their future. Their children's future. He... he would make it so. Had to.

He put his entire concentration now on the task ahead. How long would the mammoth maneuver take?

Scott, like every schoolboy, knew a beam of light from sun to earth took some eight minutes to make the trip.

He wasn't exactly a beam of light, not any more anyway, with his speed now showing well below C.

So the time... he calculated mentally maybe something like an hour at the outside. A lot depended on how quickly he lost speed once he had made the turn and pulled away from the sun's gravity... if that giant fist of force did not shake his ship apart first. He threw the outline of the problem at the computer and got a kind of shaky confirmation from it.

So... so there was, theoretically at least, still time to make his rendezvous with Jolene. The sun could be his undoing, of course. If it didn't shake him to death, it yet might fry him to a crisp.

He positioned himself inside the cocoon on the rubberized deck designed to look like a Victorian carpet. He knew he would need every unbreakable one of the thousands of miniscule straps that held his molded second skin tight now against his own... knew he would need every advantage of what the NASA engineers called the "immaculate suspension" to protect him from what was coming. From what could be worse than breaking the lightspeed barrier.

In a matter of minutes the whole universe had turned to one gigantic yellow glare, a glare so bright even the special glasses he had donned felt like they were going to melt on his face. He wondered if the computer had gotten his message right. As close to the sun as human endurance would permit, he had instructed it, and that at the end of a long slow lazy S turn. Hah, that was a laugh. He was still at ninety per cent of C, he saw on the little remote monitor at the base of the cocoon.

He felt the final turn as the ship left the enormous yellow glare gradually behind him, heard the ship's cooling system laboring as it never had before, felt a layer of sweat form all over his body.

Then it started.

The war against the sun's gravity. Tug of war might be more accurate. The ship began to shimmy and shake under the opposing forces. His enormous forward speed wanted to take him towards earth while the equally enormous gravity of the sun behind sought to stop him in his tracks, suck him backwards and then swallow him whole.

The war was fought out along the lines of his own body. He watched the G meter rise as the long last part of the turn was completed... seven, eight, nine... Then it started to drop along with his speed.

The speed dropped so precipitously now, Scott wondered if he was going to make it out of the well of the sun's gravity, after all.

Maybe he would end up yet as a solar shish-ke-bab.

The computer told him otherwise.

He was now only thirty million miles from earth and his speed was slowed to one tenth the speed of light. Thirty minutes to earth at something like a million miles per minute. Slightly less time to the moon.

Once there, his retro rockets should take care of the rest.

And the time... it was all right. He was going to make it.

He got out of the cocoon, ran his fingers through his hair, seated himself on the sofa. He punched in the program for the moon landing. He got up, went to the refrigerator, reached in. He twisted the top off a frigid beer, gulped it all down, immediately reached in for another.

Gradually it came to him what he had to do next.

Two things.

First he threw on the Stealth control. The NASA engineers had factored it in almost at the last minute as a kind of afterthought. It was supposed to be for emergency alien evasion in case such an unlooked for eventuality might somehow arise. The irony was Scott would now use it to hide himself from the NASA people at the Moon Base. The Stealth made his ship invisible to radar and almost invisible to the naked eye.

Second, he then worked with the computer to redefine his landing point. That was so he would not land on the moon's spaceport itself, but behind the mountain peak to one side, using the vertical landing capacity of the ship to fit into the somewhat narrower space...

****

There she was. Jolene.

Standing over by the big observation window looking out on the launch site.

Scott could hardly believe it.

His heart pounded, and he had to fight off a sudden spell of dizziness.

He looked warily around.

Apparently he had beaten the long-haired, bearded man. But the time was short. He had to act fast.

In three strides he was across the room, putting his hands on Jolene's shoulders, whirling her around...

Her tone showed joy, surprise and desperation all at once. "Scott! Thank God! How did you get up here? I tried to get down to the launch site to see you, but they wouldn't let me in! To tell you, I changed my mind. Or made it up. I choose you... my heart... my heart always did, but I was trying to be fair... now I don't care if it's fair or not, even though it hurts Jack. I love you. Only you. And I don't care about the rest."

Scott took her in his arms. "I don't care, either. I'm not going on the mission. Well, I am... but I'm not. I'm going to be with you. I'll explain later."

He held her for a moment, feeling her warm breath behind his right ear, holding her tight, tight, tighter than tight. Letting the perfume of her dark hair intoxicate him...

Then he bent her backwards and kissed her as hard as he ever imagined he would do throughout all the cold and lonely caves of space he had found his way back from.

Thoughts like lightning flitted across the dark fields of passion that seemed to have taken over his mind. The government might want him if they knew he were here, want an explanation, want to pick his brains. Which would be all right when that time came. Which was not yet. Definitely not yet.

When? *Maybe when Jack came back, his friend, the Flash. Not just the Flash now... Magellan...*

That would be the time to tell the whole story.

Meanwhile he would adopt another identity, another name. How could he make a living? It came to him that anyone who knew in advance of the great World Stock Market Crash of 2037 would have to be retarded not to be able to prosper from that fact alone.

About the other event, the catastrophic event that eventually must happen to earth...

Well, future events were not written in stone. His very presence here was testament to that.

With his foreknowledge of that terrible catastrophe, there had to be a way to change that, too. And he silently pledged himself even as as he lolled in the Elysian fields of Jolene's kiss, to find that way.

At last the kiss came to an end.

Jolene stepped back, a hand on each of his arms, and looked up at him.

"Scott, you know something? I like it and I suggest you keep it. At least until our children grow up."

"What? Keep what?" He suddenly remembered that what he had to keep was an eye out, looked warily around for his enemy.

"The long hair. And the beard, too. They make you look... exotic." She gave him her million dollar smile.

Comprehension came at Scott indirectly at first, like sunlight coming at an angle through latticework, leaving everything dappled with more shadow than sun.

Sure, he thought. An exotic look. So why not an exotic name, too? Let's see. Maybe some kind of rough anagram...

The name did not come to him at once, either. But he knew eventually it would.

—THE END—

* * * * * * *

Some Surprises...

First of all, Robert D'Artagnan is a pet pen name of mine.

But more importantly...

The Speed of Light... a subject that has and will keep coming up in this book, necessarily.

Einstein and Einstein's formula say that C can not be exceeded... ever, under any circumstances.

If it were, Relativity would have to fold its tent and steal silently away.

So... what does AAA Einstein say about that?

I already said it in the story. At least I think I did.

Before the sound barrier was broken, a lot of scientists said it never could be... or if it ever was, the results would be disastrous... total disintegration.

The sound barrier was broken and everything hung together, although under greater stress, of course.

If we take into account this factor... that an element of the Universal Medium is the Fifth Dimension... then that opens a door for us, just as it did for Scott Diamond in the story.

That is... while it is entirely possible that the Speed of Light can not be broken through, being the limiting speed of the Universe and also being, as I have pointed out, the ultimate speed of transmission of the Universal Medium... still, what happened in the story could very well really happen in the real Universe.

How?

As described.

At the point of actually obtaining the Speed of Light, but just short of it, the 5th Dimension itself may come into play. While the version of the 5th Dimension we have in Our Universe may be just an incredibly tiny rolled up cylindrical version... at proximity to attaining the Speed of Light it might open up like some kind of space flower. Flare out into something more like what it probably is in The Greater Universe that surrounds us. Into a corridor. A timeless Time Corridor that once attained can not only take you to other places... maybe anyplace in Our Universe... but also to any TIME in Our Universe.

That, as you have seen, is what happens in the story.

But if you don't like that particular explanation, I have another... which I think is something of a jewel in itself, as a concept.

The alternative explanation for what happens is this...

While the spaceship might have almost infinite Mass at that point, its directional length would shrink to almost zero. Which might permit it somehow by some twist, some physical *Abracadabra* to enter the tiny, microscopic cylindrically compressed 5th Dimension.

By the way...

Einstein himself was offered the idea of the 5th Dimension by a Mathematician named Kaluza. And it gave him a way to incorporate Electromagnetism and Gravity into what was his dream of accomplishing... a Unified Field Theory...

The year was 1919 when Einstein received a letter from Theodor Kaluza, who was a Mathematician at the University of Konigsberg. He had noticed something very special about Einstein's theories and had come up with an idea. If another spatial dimension could be factored in to Einstein's equations, then something magnificent became possible... Einstein's dream seemed attainable.

Because...

With the addition of the new dimension, the apparently different physical forces of Electromagnetism and Gravity could... mathematically, at least... be reduced to Gravity alone!

The Unified Field!

Then in 1926 the Swedish Mathematician Oskar Klein came up with something else. Something very important and apropos...

He said this...

Yes, the extra dimension might conceivably be the mediator between electromagnetism and Gravity. But if it were, nature would require that the extra dimension must have an extremely tiny extent... something like 10 to the exponential power of MINUS 33 centimeters!

What that meant was that the extra dimension would be so infinitesimally tiny... that it would be both invisible and undetectable.

Beyond human power to observe.

Of course that sounds to me like the specifications I have assigned to the Universal Medium... including the 5th Dimension contained therein.

I like to think of it like this... as a kind of Underworld... BELOW all our powers of observation, but ironically POWERING AND CONTROLLING everything that goes on in the Universe.

Lest there be any doubt about the possible import of these two mathematical innovators... Mr. Kaluza and Mr. Klein...

Look at this scenario...

If you started by recognizing the existence of the Universal Medium... Then you punched into all your equations Oskar Klein's mathematics concerning Kaluza's 5th Dimension... you might very well work your way towards achieving a beautiful Unified

Field of Gravity and Electromagnetism and perhaps even all the other known forces of the Universe.

Not such an idle dream. Not at all.

It was, in fact, Einstein's Dream. On the point of becoming Reality.

But Einstein... for whatever reasons... though he at first considered them both... at last turned a blind eye and a deaf ear to these original thinkers...

Well, nobody's perfect.

So...

What might have been is very different from what was...

As to my story...

I put it in here only for dramatic relief... from what may have been the understandable shock of realizing that the basic structure of the Universe may be both simple and accessible. So simple that (a) even a AAA Einstein can explain it, and have fun doing it... and (b) the reading audience can, worst case, at least understand it... and best, enjoy it just like I do.

Hopefully we are now past that, so we can go on to something else. *All right, Superman, you jumped over a colossal building. That was then, this is now. What's next?*

And, now that I think of it... what IS next?

****

Well, there is history, and elaboration, and speculation. But all that is saved for later to keep everything here up front as simple as can be, strictly AAA Einstein... the beyond part we will take up somewhere along the way.

The fundamentals we have covered in just a very few pages. Let's review them... the origin of the Universe... its basic shape... what Gravity is and Time... both governed by the Universal Medium... and... that's it.

Hopefully...

So why did Einstein make it so complicated? Well, to get from Kansas City to Lawrence, Kansas, you can drive forty miles. Or... you could go 24,960 miles in the opposite direction.

I think Einstein made the wrong choice. Instead of recognizing the validity of the Universal Medium, he brought in new and difficult factors... principally the idea of Curved Space with all its concomitant complex geometry and difficult mathematical equations.

And then how did he correlate that with Quantum Mechanics? He could not...

While the UM... the Universal Medium... heals all those unnecessary wounds. At least, potentially so.

****

Perhaps I should say a few things here about the Speed of Light, some of which were shown in the story...

Question. Does the Speed of Light vary?

Absolutely. A word I have chosen carefully.

Light travels slower in water than it does in air. Slower in air than it does in a vacuum. In a vacuum? Wait a minute, a wave cannot propagate itself in a vacuum... even though modern day science likes to pretend it can. Let's modify that, and say in the Universal Medium.

Light travels at the maximum speed permitted by *whatever medium it is traveling through*. Is that too simple for anybody? You and I aren't too different from light waves in that... we slog through mud, but can speed over firm, grassy turf.

Molasses slows the light waves down just like they would you, had you yourself to navigate through a knee-deep field of the sticky stuff.

Light also travels more slowly when under the influence of heavy gravitation. Inside a Black Hole it can stop completely, become frozen just like Time.

So in a sense, Light and Time are twins. At least insofar as the way they both react to Mass, which is really not Mass itself, but the concentration of Gravitational Force inside and around Mass through the varying concentrations of the Universal Medium.

To get to the bottom of things, it is The UM, the Universal Medium or Field of Force, that controls and regulates all the workings of the Universe. Once this is admitted everything tends to become simple and apparent.

But we said Light and Time are twins in that they react equally to the presence of Mass.

And understanding that, it is easy to understand why Light between astronomical bodies is often measured as being a constant 186,282 miles per second. First of all that is the NORMAL transmitting speed of the UM. And in those cases where light is slowed by concentrations of the Universal Medium, either by Mass or accelerated

Mass, Time is also slowed in an exactly proportional way.

The Time-Light Twins.

Heavy gravitation can make Light and Time slow just as if both were trying to make their way through a sea of molasses.

Say you are on some gigantic planet 20 times larger than the sun. And say you have survived the gravitational stress on your body...

Then, while Light in its local travel may be slowed you will never notice it. Why not? Because your clocks run slower... your time has slowed... permitting light to be MEASURED locally as 186,282 miles per second, just as it is in most places under most circumstances.

Time and the Speed of Light are both variables in our Universe... even though the original Einstein and his interpreters have made a point of promoting Light as a Universal Constant in regard to its speed.

But Light is a variable... as is Time... if seen from any outside vantage point. If you could stand outside our Universe and observe Light and Time, you would see this... would see them both varying considerably depending on the circumstances in any particular part of the Universe...

Yes, the following is true...

Light and Time are variables that tend to exactly compensate for each other. Strange? No, actually not.

Actually no mystery there.

Since each is governed by... the common factor of the local distribution of the Universal Medium.

This is another way of saying again that it is the *distribution of the UM field of Force* that regulates and controls everything in our Universe.

In fact, if you want me to bring everything down to its simplest terms... which I am sure you do...

Then...

I can say this...

Gravity, Time and the Speed of Light are all three ruled by the same King. The Concentration or lack of it... the distribution... of the field of force of the Universal Medium.

A more simple principle could hardly have been expected concerning the workings of the Universe.

And so we tend to find that the physical laws that apply on earth would also apply in any other part of the Universe... because of this basic unifying factor that is the combination of the Time-Light twins working under the influence of the Universal Medium and its varying concentrations. This to a large extent is what holds the Universe together and makes things run under general control, at least on the Macro scale.

Question...

Why could not Einstein himself have told us all this, instead of making things so unnecessarily complicated?

My answer...

I think he would have liked to... to talk to us person-to-person... but it is difficult to communicate from within a straitjacket. While this straitjacket... this rigorous protocol, is admittedly in many ways a great strength of scientific procedure... this same protocol hardly permits plain simple talk accessible to the great majority of people.

For that you have to turn to... AAA Einstein... who comes not to destroy science but merely to try to open up some of its many closed doors and let fresh air circulate... and let some scientists break out of their straitjackets.

The time for that particular bridge to be put in place may well be now.

****

It may be too much, a case of overkill. But I feel almost obligated to introduce at this point still another explanation.

Call it my idiosyncrasy. But still, we ARE talking about the workings of the Universe here...

So here goes...

The Speed of Light... Gravity... Time. Pretty important things to talk about. And what I want to do, if I haven't done so already, is to show in a perfectly clear way how they are interlinked and how they work.

First, imagine Light travelling across the Universe on a kind of track... that track is the Universal Medium.

The Density of the Universal Medium is not constant across the Universe. It is one density in so-called empty space far from any astronomical bodies. And quite another near those bodies, and their gravitational fields.

I like to imagine the Universal Medium this way.

Picture it as a network of billions and billions of interstices. If you have that in your mind in 2D, now project it outwards in 3D... and you begin to have a workable mental picture of how the UM may work.

Now Gravity rides upon this network and directly influences its structure, its density. Gravity is generated from the center of Mass in a basically spherical pattern, radiating its force outwards equally in all directions from the center of Mass.

Gravity contracts the Universal Medium's 3D network. Take Planet Earth. There are more lines of force of the Universal Medium at the center of the Earth than anywhere else in its vicinity. From that center outwards you have a series of concentric spheres, each one of which is less dense in lines of force than its neighbor closer to the center of Earth.

Gravity's force brings the interstices of the Universal Medium network closer together. Makes it more dense. Thus, the progress of Light is slowed as compared to the less dense background of empty space.

Now here's the thing, remember... no matter where you are... the Speed of Light will not be MEASURED to be different. Why not? Because Time itself will also be different... in exactly the same proportion. Because what is Time? Just the rate of all atomic movement including vibration... which has been also moderated by the particular local Concentration of the Universal Medium.

So...

There it is still once more in simple language.

AAA Einstein language.

How the Speed of Light and Gravity and Time are interlinked. They are all three keyed to the concentration of the omnipresent Universal Medium. Once this common factor, this central controlling influence of Our Universe is understood... then the picture of Our Universe becomes much clearer than perhaps it ever has been before.

Anyone can visualize it. Even I, who sometimes find it difficult to comprehend maps, even those from AAA...

* * * * * * *

Einstein correctly keyed all physical reactions in the Universe to the Speed of Light.

This was a major accomplishment, a tremendous insight.

But...

He did not take THAT LAST STEP...

To show that the Speed of Light itself was IN ITS TURN keyed to something EVEN MORE BASIC in the construction of the Universe. That something more basic is the presence of the Universal Medium everywhere.

Had he done that, I would not now be writing this book, because I would have understood it all from Professor Einstein.

But here I am plugging away...

Why? Because...

That one step further is important, because with it, now we have an insight into how the entire Universe works.

The Speed of Light... its concept... is changed.

It becomes no longer a kind of arbitrary, mystical guide to everything. Now we understand what is behind Einstein's unexplained mystique, perhaps not understood by Einstein himself...

Now...

Now we can now see HOW the Speed of Light constantly varies, yet usually comes up being MEASURED as a constant... because the Speed of Light depends on THE TEXTURE of the invisible underlayer of the part of the Universe it is travelling through. As does Time.

So...

We can say other things...

We can say that the Speed of Light in a vacuum is... the maximum known transmitting speed of the Universal Medium in which it travels.

So in our hopefully deeper view, we are saying that instead of everything in the Universe being keyed to the Speed of Light... everything in the Universe is instead keyed to THE DENSITY OF THE UNIVERSAL MEDIUM.

This is quite a change. It does, in fact, change everything. It is what intellectuals... none of them known to this AAA Einstein... might call a Paradigm Shift. I only understand that term in so far and if it means... time to re-examine EVERYTHING.

****

If all that is too simple for anyone... in the sense that you say... it can't be true, because even I can understand it... then you are in the same boat with me.

That is why I am trying to make myself into AAA Einstein. And I would welcome other passengers to come aboard, of course...

Randall Barron

Another way of saying what has already been said might be this...

It is the distribution of Mass in Our Universe that IN A PRIMARY WAY determines most physical processes. Gravity... Time Flow... and the Speed of Light (actual as versus measured). And the GOVERNING ENTITY that CARRIES OUT all these complex instructions, is the Universal Medium.

If everything is keyed to the UM, as influenced by the presence of Mass, as it properly should be, then we have an orchestra working together in harmony... an orchestra which can and does produce what we might call a kind of Music of the Spheres...

****

What I have tried to do in a few pages, is give an overview of how the Universe was created and how it continues to exist according to a few simple elements.

At the same time let me make something clear. This is not a children's book, in spite of the fact that the sentences may be short and the ideas simple. Nor is it an attempt to popularize existing knowledge or perceived knowledge in the scientific community.

Not at all.

It IS an attempt to postulate something original and new. What I hope to be a valid interpretation of the basic physical laws of our world and Universe. Based in general on Einstein, yes, but as you have seen and will see, somewhat different, too.

In many ways it is a step... at least... beyond...

It is...

What I like to think of as AAA Einstein... the kind of explanation you might theoretically get from, say, your local automobile club or your local auto mechanic... nothing all that complex, although it may contain a lot of distilled and concentrated knowledge, which is definitely very useful.

My modifications of Einstein's theories and my ideas on Cosmology may also solve some other important Scientific Mysteries. The Mystery of the Missing Mass in the Universe. And the Mystery of the Missing Energy...

More later on those.

****

I think I have accomplished the purpose of simple and visualizable explanations, hopefully based on things even more fundamental than Einstein's principles... to my satisfaction at least, up front as advertised by the title and over the space of a few pages. The remainder of the book goes into history and the projected future, and all kinds of musings and speculations.

But everything is built upon and around these first few pages. Nothing fundamental will be changed.

It is my hope, then, that my Redneck Science will have caught on with at least the Redneck element of society. The auto mechanic or school teacher or supermarket clerk or retired person from any occupation... who always wanted to understand science but was dismayed by all the esoteric terms and mathematics and, above all, a kind of attitude... an attitude that said more clearly than words... this is not for you... this is restricted territory... this is the area 51 of science... reserved only for us specially gifted people.

If I have overcome that barrier, I am more than content.

Now I can elaborate with speculation and additional information that may take us... anywhere.

Sometimes, I admit, I even feel the presence of Albert himself... looking over my right shoulder.

He is not being protective of himself or censorious. On the contrary...

Go for it, he seems to be saying. Just as I already have against the entire ponderous Scientific Establishment.

* * * * * * *

Genesis

In the Beginning...

What really happened?

And what is really happening now in the real Universe outside of all theory and scientific texts?

Is there any way we, the ordinary people, can tune in on this and really understand it?

Maybe there is a way. At least that is what this book is about. You see, I think a wrong turn was taken in the history of science... a detour that has led us away from what the true and simple picture of the Universe could and should be. That wrong turn was taken by none other than the greatest genius science has ever produced... the nonpareil Albert Einstein.

He thought his greatest blunder was in not recognizing from the beginning that the Universe might be in a state of expansion... and as a result at the end introducing into his formulas what he called the Cosmological Constant to balance things out... so as to portray a Universe in a stable state.

But no, that was not it. Not his greatest blunder. What that greatest blunder was, how and why he made it, and how it ended up muddying the scientific concept of the Universe... that is what will be taken up in the remaining context of this book.

Here I will only say by way of introduction, that what could have been a simple, clear inspiring picture of the Universe... one that all of us, including even primary school children, could embrace in its outlines and essence with at least a competent understanding of the overall picture... was somehow ambushed and extinguished.

Instead we got theories and explanations so intricately difficult that the popular press of the time said that possibly only twelve persons in the entire world could understand them.

I do not imply here for a moment that the physics and the mathematics of the Universe are simple enough for everyone to comprehend and be able to work with. No, but...

But their principles are, or should be. Those principles should not be restricted to a total comprehending population of just twelve people.

And so this book. To point out Einstein's errors... if you can get by the colossal effrontery of anyone who challenges the great master... and try to draw a truer picture of the real everyday Universe we live in.

Hopefully, in the end, we will all have a truer idea of the answers to some perennially unanswered questions. Such as...

What Gravity really is. Not just what it DOES, but what its essence is, and how it works.

Why it is that Acceleration and Gravity produce similar effects. Sometimes.

Why Time is a variable.

How Light constantly varies in speed in its travels across the Universe, but how, yes, we usually... but not always... MEASURE it as being the constant quantity C, and why that is.

How sometimes the medium really is the message.

How Light itself seems to defy Einstein's theories.

Above all how there is one simple, neglected element that can bring everything together... can unify the microscopic and macroscopic worlds, put them both under the same umbrella. This was an element that Einstein chose to ignore... to his detriment and our peril.

And had he not ignored it, he might well have achieved in his last thirty years of life, his dream... the fabled and never attained Unified Field Theory.

And help me out, please. Because...

A book like this is too important for me to be able to achieve its objectives too straightforwardly. I may have to tack here and reef sail there, go sideways at times and even reverse course if I really wish to advance eventually, which I certainly do.

Perhaps I can even give Science a literary veneer. Introduce a little French Impressionism into it as if it were a painting. I don't know. I am writing this before I undertake my daunting task. Dreams and ideas do not always coalesce as AAA Einstein would like.

But my intuition tells me I will have to feel my way to get across what I intuit to be true. I will try to go with that intuition and go with whatever methodology occurs to me in the process. I know this is not scientific procedure. But perhaps if Albert had gone more along with his intuitive feelings, we would not be now in this dark *Cul-de-sac*.

In the end that is what I propose to try to do.

Let in the light over an unnecessarily dark area and let the basic genius and insight of Albert Einstein at the end shine forth brighter than ever before.

Wish me luck.

****

Past History, Continuing Mystery

The nineteenth century.

Do you not long for it?

Because...

Because I do.

Things were simpler then. We were nearer to God, and farther from Radical Terrorists. For another thing we could understand the workings of the Universe through the theories of our scientists.

The most important figure giving us that insight into the workings of the Universe was Isaac Newton.

And what a figure he was.

The first to try to put a habitation and a name on Gravity.

Gravity lived in Mass. That was what it was, Gravity. Mass.

Its attraction was legislated in a very civilized and simple, easily understandable way. The more Mass the more gravitational attraction. Everyone could understand that.

Proximity was important.

The closer together two massive bodies, the greater the attraction. You just multiplied the masses together and divided by the distance they were apart... well, the distance they were apart multiplied by itself, or squared. That wasn't hard.

God was in his heaven, Isaac was contemplating the fall of apples in his garden... we understood him and his genius... all was right with the world.

We began to get an idea that God was very like a Watchmaker. He wound up the Universe and let it run like a great clock. It was Law and Order on a colossal, universal scale. Fair, just, and above all, accessible to the understanding of everyone, at least in its principles... and yes, even in its mathematics, that were not so daunting as to be totally exclusive.

No one ever said that Isaac Newton could only be understood by twelve persons in the world...

His picture of the Universe was encouraging to the idea of human progress. We could work, study and learn and eventually hope to achieve understanding of the Universe. Of the Mind of God, perhaps...

But wait...

The picture was not totally rosy, nor even complete...

There were still unanswered questions.

What WAS Gravity, really?

A universal attraction between masses, yes. That was clear.

But how did it work?

Two masses, say the moon and the Earth... separated by a quarter million miles or so.

How did Gravity bind them together?

How did they reach out to each other through empty space?

And how long did it take?

For this force to take effect...

Was it instantaneous?

Science took a stand on some of those issues. Yes, it was instantaneous, the force of Gravity, and yes, it acted at a distance.

All right.

But what was it?

That...

That was a question left unanswered.

Not only unanswered but never even voiced, never asked.

Perhaps because the lack of an answer could have been embarrassing to the Scientific Establishment.

Nothing new there.

Still, some day the unspoken question had to be answered, if science were to forge ahead, to progress beyond Sir Isaac's Great Leap Forward...

****

Enter the 20th Century...

The next person, the next Giant of Science, to address the problem of Gravity was Albert Einstein.

He began with an observation. An observation based on one of his thought experiments.

He used his imagination.

What if...

What if you were in a space ship and that space ship were accelerating?

To you it would feel that Gravity was increasing.

And so Einstein puzzled over this congruence... that Gravity and Acceleration could apparently produce identical effects.

What did that mean?

Well, there is evidence to suggest that Einstein never was able to answer that question satisfactorily to himself.

He knew he had a tiger by the tail.

Something of extreme importance. A clue to the workings of the Universe.

But what, really, DID it mean?

****

If Albert Einstein would have been able to incisively resolve that question, we would not be in the metaphorical dark we are today.

He might have long ago formulated his Unified Field Theory... and the world would understand in principle how everything works.

But he WASN'T able to resolve the question.

* * * * * * * *

His frustration must have been extreme.

The tiger was at the point of devouring him.

He had to come up with something.

He did.

The concept of Curved Space. That that was what Gravity did. It deformed and distorted the contiguous space around any Mass.

From that moment forward he was on a wrong road.

At no point along that increasingly lonely road was there in view even the ghost of his vaunted dream... the Unified Field.

Not only that, it got in the way, the concept of Curved Space.

He could not get together through it the idea of how Gravity and Acceleration... two separate phenomena... could produce the same effects.

Curved Space blinded him to what might have otherwise seemed obvious even to a lesser mind...

****

Science made great progress in the Twentieth Century. The effect was something sweeping, something revolutionary. In fact, what happened might well be compared to the French Revolution. While the ideals were high, the actual results were too often awash in a sea of blood to offer much satisfaction to either intellect or soul...

In the French Revolution it might be said the baby was thrown out with the blood bath... and so, in a way, too, was it with the Science Revolution.

The French Revolution wanted nothing at all to do with the decadent French Aristocracy that, so it was thought, had caused all their problems.

Twentieth Century science wanted nothing much to do with Nineteenth Century science, either.

Phlogiston... it was nice to get rid of that.

But Sir Isaac Newton? Careful there... his theory of Gravitation had worked pretty well... and above all he had the basic idea that Gravity was associated with Mass, and that proximity had a lot to do with the attractive process... certainly valid basic ideas.

And there was another idea of Nineteenth Century Science... one that was very early tossed aside as invalid and worthless...

The baby...

What on earth am I talking about here?

Well... a lot of very valuable work had been done on electromagnetic phenomena. Basic to the theories proposed was the idea that these electromagnetic phenomena were wave motions. And a wave by definition had to have some kind of transmitting medium to move from point A to point B. On earth the transmitting mediums might be air or water or the earth itself. But what about in outer Space? Light obviously travelled through it. We received Light from the sun, the moon, the stars...

So how was that done?

Science had to come up with some kind of explanation. And they did...

The answer was... the Ether.

What was the Ether? Well, it was an invisible, undetectable substance that pervaded the entire universe and served as a transmitting medium.

It sounded fantastic, and a little like the concept of Phlogiston.

But it worked very well as an explanation for electromagnetic phenomena and their propagation across eons of empty space.

Twentieth Century Science could not wait to get rid of it.

It was mistaken... unnecessary... an obstacle to progress.

That is what was said.

The Michelson-Morley experiment of 1905 became the focus of the drive to get rid of the concept of the Ether.

That was a surprise to Michelson and to Morley. The announced object of their experiment was only one... to get a still more accurate measurement of the speed of light.

This they accomplished. But other scientists quickly placed a different interpretation on the event. They said it had proved the Ether did not exist.

Just how it did that is a little hard to say.

Because basically all Michelson-Morley did was to flash light rays in four directions for equal distances (very short distances, less than thirty feet) and bounce them back again to their measuring device. That measuring device showed there was no difference in light speed directionally.

?????????

That was supposed to disprove the existence of the Ether?

But the idea swept the scientific world. The Ether was dead. Louis the Fourteenth had been guillotined. His head had rolled and he was no more. Now the Revolution could progress.

New theories could be put in place. Robespierre's instead of the deceased King's.

In the case of the Scientific Revolution, things looked decidedly more encouraging than had been the case in the French Revolution.

There was Einstein. His new theories were brilliant, beyond question. There was resistance to them at first, yes... but very quickly as experiment seemed to verify his formulas, he was accepted and became the new Administrator of Scientific Truth. Isaac Newton became almost another casualty of the new regime.

Einstein deserved to rule. He was a true genius, with deep insights into the inner workings of the Universe, that went even beyond those of Sir Isaac.

But...

No one is perfect. Not even Albert Einstein.

His biggest mistake, as I have hinted earlier, came now...

Undoubtedly influenced by Michelson-Morley, he threw out the baby.

He never mentioned the Ether in his exposition of his theories.

He ignored its possible existence.

Not that... and note this, please... not that he ever denied the existence of the Ether. He did not. In fact, we have a statement that shows he was not opposed to the idea of its existence.

And had he recognized it in his equations, Science would be in a far different, and I believe much more advanced state at this moment.

Instead, he interposed the idea of Curved Space.

This... this has been a concept very difficult to work with. How do you curve something that has no observable physical existence? There are very serious doubts to this basic flaw in his otherwise ingenious concepts of the Universe.

And Einstein and Twentieth Century Science have left us with various Unsolved Mysteries.

Just to list a few...

1. Why there is very little correlation between Quantum Mechanics, which attempts to describe the workings of the subatomic world, and Einstein's theories of the everyday macroscopic world.

2. How the force of gravity is transmitted across eons of empty space, as we know it is, with no intervening medium. Isaac Newton did not resolve this enigma... neither did Albert Einstein.

3. How suddenly in empty space there appear spontaneously and momentarily particles and energy emanations... apparently from nothing. A phenomenon very often observed in modern times.

4. How pairs of photons, or basic light energy packets... even though separated by great

distances... react to each other's change of spin by instantaneously changing spin also... a reaction that according to Einstein should be impossible.

5. Why it is impossible to determine at one and the same time both the position and momentum of an electron inside an atom...

****

Back to the Future...

What I am going to suggest here is something very radical in itself, and something not likely to receive any kind of enthusiastic reception on the part of any scientist. The explanation of that, supposing I am right and they are wrong, is simple enough. Science builds on its own base... there is no need to re-invent the wheel, or disprove once again what has already been disproved.

The Ether and Anathema... they are both about the same.

Or we could take the example of phlogiston... a known blind alley and mistaken concept that only led to confusion. In the late 17th century phlogiston was a theoretical chemical supposed responsible for the quality of flammability in all substances. But then Joseph Priestley discovered oxygen and Antoine Lavoisier proved it was the key element in combustion... and as a result phlogiston as a concept was dead and buried by the year 1800.

But what I suggest to you here is there has been a colossal turnaround and mix-up.

That the Ether is a correct and valid concept. That it is the concept of Curved Space that must analogously assume the role of phlogiston.

That Einstein was wrong in ever buying into that concept. It put him into a *cul-de-sac* from which he struggled... unsuccessfully... to escape the rest of his life.

And conversely, that had Einstein retained the concept of the Ether...

Well, for one thing I would not be writing this book.

You would not be reading it.

There would be no need.

The modern scientific mysteries I have suggested above would in all probability no longer exist.

Everything hunky-dory. Explained. *Tutti contenti.*

Your fourth grade sons and daughters understanding perfectly well the Principles of Relativity.

But of course it is not so.

The enigmas of modern science remain.

I have said a lot about the Ether... about what it is and how it works. Though I will replace it, as I said previously, with a new term, and a modified concept... the Universal Medium. And then I still have a lot to say about Gravity, and how Mass and Time can vary.

Basically, all this will be just Einstein revisited. AAA Einstein... accessible, atomic, and artistic.

The way it should have been from the beginning.

But it takes the Ether to put things right. Not necessarily exactly the same Ether of Nineteenth Century Science... because the UM contains some added elements they perhaps had not dreamed of... but still, basically the same. A hidden truth of progress is that sometimes you have to take a step backward to advance...

Just to begin with...

How would the above mentioned Unsolved Mysteries of modern science be resolved if my more modern version of the Ether, the Universal Medium, were taken into account as a real part of our Universe?

Well...

Let's take them up, one by one.

1. Little correlation between the microscopic world of Quantum Theory and the macroscopic world we see everyday...

The answer is that the Universal Medium is the intermediary between the two... the UM is the Great Energy and Matter Interchange that works on both levels. It fills in the blanks between one world and the other.

2. Gravity can at last be understood... something never possible in the deepest sense with either Isaac Newton or Albert Einstein. At last we have a transmitting medium... one that permeates just as much the sub-microscopic world as it does the everyday macroscopic world. Not only that, but as we have seen, Gravity is an integral part of the Universal Medium itself. For the first time we may catch a glimpse of what Gravity really is...

3. The energy and matter emanations in the middle of empty space... the apparently spontaneous generation of "virtual" particles, what amounts to almost ghostly apparitions... for the first time, can be

explained... if we realize the Universal Medium exists everywhere... and is the Ultimate exchange medium of the Universe. Processes are constantly going on within it to maintain the equilibrium of the Universe and keep everything under a system of strict law and order.

4. The observed phenomenon of paired photons or energy packets which always maintain opposite spins... even though separated by light years. Something patently impossible with today's scientific concepts. The answer would be the UM connection. As to how such a connection could be instantaneous, I will save that for more advanced reading in this book...

5. The Uncertainty Principle... so well established in Quantum Physics... becomes understandable if we realize the Universal Medium is a vast well of resources of fundamental parts... an Energy and Matter

Exchanger... and that electrons at any time can change from dissipated cloud to virtual particle and back again... that this is only possible through the Universal Medium with its exchange capabilities... that contact always tends to concretize the electron, while in its lack the electron tends to become a cloud... much as certain salts applied to chemical mixes tend to precipitate out its basic constituents, while in their absence the chemical mix remains a mix.

****

I want to write about a great many things in this small book, and while everything being interconnected in the Universe is, I believe, a greatly useful and correct concept which will further our understanding... at the same time, that very interconnection makes it difficult to find a starting point and a straightforward developmental line...

On the contrary, things tend to move in a circular and globular direction, imitating the reality that I am trying to describe in terms simpler than those Einstein himself used. Simpler, and different, too... since Einstein in my opinion definitely took a wrong turn when he accepted the Curved Space system of mathematics and geometry to further explicate his ingenious theories.

Curved Space merely confabulates things, throws them not only into a cocked hat, but exponentially complicates everything and results in a picture of the Universe that is... that is, what?... if it is a picture at all, pardon me Albert, it is a very ugly picture.

You have to visualize a space which is a kind of construction. A solid framework of emptiness which must alter at any time according to changing conditions. I have called it ironmongering in the sky, because that is what it appears to be to me.

And I think it is definitely mistaken, and not at all in step with reality.

I have given you a better picture, one I think is also more valid. That picture involves the Universal Medium, which is far easier to work with than Curved Space.

I can't help but think that Einstein must have toyed with the idea himself. After all, the Ether was an accepted part of the scientific Universe at the time he formulated his Special Theory of Relativity. And if he looked at it objectively, it certainly could have accommodated his theories perfectly. So why he did not incorporate it into those theories is something of a mystery in itself.

Maybe it was the pressure of other scientists so eager to declare that the Michelson-Morley experiment had proved the Ether did not exist.

I don't believe it proved any such thing.

If Michelson and Morley believed that, they were certainly very reticent about expressing any such idea. They only wanted to get a more exact measurement of the speed of light, and that they accomplished.

Did they accomplish something else? To disprove the existence of the Ether?

That was a conclusion in the minds of other scientists, but perhaps a mistaken one.

How could the experiment prove that?

The reasons given are not very clear.

A lot of talk about the Ether Wind and how it should affect the speed of light... but apparently did not.

I don't follow that...

Michelson and Morley generated light and projected it in the four cardinal directions and found there was no variation in measured speed.

So?

It seems a major *non sequitur*. Something that does not follow...

Can we compare the situation of the earth moving through space with something as prosaic and nearer to home as this... a boat moving through the water of a river?

If from the boat we send out sound signals... sonar... in the four cardinal directions, then... Do we expect some of those signals to be returned at different times because of our forward progress through the water?

I don't think so. Not if what the signals are being reflected from are all equidistant from the boat.

The speed of sound through the water is dictated only by two things. The natural propagating energy of sound waves and the medium through which those sound waves pass... in this case, water.

The current of the water is not going to change the speed of sound in water, nor should we expect it to.

There is a Doppler effect, yes. The pitch of the sound in the forward direction may be higher, the pitch of the sound in the aft direction lower...

but the velocity is the same. This is what Sonar is based on.

Just as we have in our expanding Universe the Red Shift from distant Galaxies receding rapidly from us. But the measured speed of that light remains the same.

Just as the pitch of a train whistle coming towards us rises, then falls as it speeds away... but the velocity of sound through air remains the same. The train because of its forward movement throws a couple of extra whistle vibrations towards us as it approaches... and a couple less as it speeds away from us... but the velocity of sound does not change as long as the medium through which it moves does not change.

So... in my opinion, Michelson and Morley proved nothing at all in relation to the Ether, neither pro nor con... nothing.

But... scientists believed they did.

Scientists believed that the existence of the Ether had been disproved.

And for Albert Einstein, that meant he could not postulate his theories depending upon a Universal Ether. And so, Curved Space... ironmongering of emptiness in the sky...

****

AAA Einstein... and Beyond...

Yes, I do want to describe my own vision of the AAA Einstein Universe... and in great detail... but in our great globular progression towards that goal, perhaps it might be best at this point to give you here the content of an article I wrote back in 1996. It sums up a lot of things in a way I don't think I could improve upon now... and it may even some day turn out to have some historical importance.

Anyway, here it is...

****

157

Time for a Second Look?

—Randall Barron

Political, economic, philosophic, and social ideas come and go, are in vogue for a while and then fade from the screen of present reality. This is a generalized parameter or profile of progress from which we can presumably learn something. The basic but not always clearly spoken or acknowledged presumption is that progress, like entropy, has only a one-way arrow...

In a general way, that is true. But can there be exceptions? I would like to discuss what I consider to be a very large exception to the general rule.

First of all, I would suggest that science is subject to the same effects of progress as is any other facet of modern civilization.

In general this is beneficial and enlightening. Progress is, indeed, the undeniable motor of our success in conquering and surviving in what too often seems a hostile and threatening environment.

But sometimes it is at least theoretically possible for science itself, even while making great strides of progress, to go down a blind alley. Take a wrong turn. And I think this may be what has happened. As a result, we find ourselves in a position a little like that of Hamlet, who lamented that "The time is out of joint."

I realize full well my ideas may seem on the one hand revolutionary, on the other so old-fashioned and out-of-date as to be laughable, and on first reading may seem crazy beyond redemption.

The consolation is that in theoretical physics those particular qualifications can mean I just might be on to something.

First, let me start by giving a little historical perspective.

Sir Isaac Newton was the first giant of modern science. He gave us gravity, and for the first time enabled us to see that the universe was connected, bound together in ways mysterious but subject to mathematical measurement.

It was a beautiful, grandiose and great achievement.

But there was a problem with it.

The problem with Sir Isaac's concept of gravity was simple enough. It was an explanation that needed an additional explanation. One that was not forthcoming.

How did gravity work its effects? Instantaneously and at a distance without a transmitting medium...

And so Science came up with an answer. The luminiferous, all penetrating, invisible and undetectable Ether, capable of transmitting light and electromagnetic waves across what looked like empty space.

All right. And <u>tutti contenti</u>.

But then came Michelson and Morley. Their experiment seemed to indicate there was no Ether. That, at least, was the conclusion drawn from it.

And Einstein. The Great Master spoke and gradually, over time, his word became law. $E=MC^2$ became the cornerstone of a new physics, a new concept of the organization and construction of the universe.

The Master's word on gravity was accepted as law, too, and verified in experiment.

Yet there was a problem there, too. At least in the spatial and visual concept Einstein entertained to back up his mathematical formulas.

He talked of warped space. That was difficult to conceive of in itself. Principally because space is considered to be empty, to have no content. If something is empty and has no content, how can it be warped? It was a question Einstein never got around to answering.

The descriptions other people gave about warped space were mind-wrenching, too. Small objects falling into larger ones and following only one certain path because that path had been forged out of empty space so there was only one channel to follow...

Difficult. Very difficult. To think of all that conceptual wrought iron in the sky where there was nothing visible...

And if you brought it down to the world of atoms and molecules, even more difficult to conceive of such iron-mongering on that microscopic level.

Quantum mechanics seemed to be a study almost completely separate from relativity. There was little to unify the Macroscopic and Microscopic worlds.

And yet the same laws had to hold for both. Didn't they?

Well, yes, mostly. And sort of. And then again, no, when you factored in the uncertainty principle. Heisenberg and Dirac and others had their own laws in their own private kingdom, not necessarily tied closely to the theory of relativity.

In some ways, the microscopic world still flaps loose, and seems sometimes to be Dante's Inferno or

Chaos Revealed rather than a search for order and unity. This in spite of enormous progress in the past thirty years towards comprehension and ultimate simplification of the workings of the sub-atomic world.

So, what is missing here? How can we indeed bring the Macroscopic universe of Einstein and the Microscopic universe of Heisenberg and Dirac and company together? Make them seem for the first time part of the same Universe...

The possible answer, as I implied previously, may seem fantastic, preposterous.

Because for one thing at first it will seem a simplistic throwback to a previously discredited concept.

Yes, Virginia, I am talking of none other than the Ether. But not quite the same Ether, either, as previously conceived of.

The Ether, after all, does not have to be static and stiff.

It can be flexible. All-penetrating, yes.

And the simple fact is, viewed thusly, it can account for all the mysterious measurements of the speed of light, and all the distortions of time and space described in Einstein's universe. Simple compression and expansion can do that.

If matter can be conceived of as a holder and concentrator of the Ether, or what I prefer to call the Universal Medium, then everything else falls into place. Matter then distorts not space per se, but rather what inhabits space. The very fine cloud of The Ether or Universal Medium that penetrates everything in the Universe. Almost all of Einstein's conclusions can hold, just that the visualization of them is different. And, yes, too, the reality itself is different.

More needs to be said of what such a theoretical Universal Medium or Ether is like.

Here are just a few observations.

It is the substance which connects everything to everything else in the Universe. As water is a kind of universal solvent, so the Universal Medium is a kind of universal energy and subatomic matter exchanger. It is the great umpire and arbiter of ten million different energy exchanges at any place in the Universe at any time...

In fact, it is hard to conceive of the Universe operating under any other principle. Because it is connected, it is a whole, and no matter what extraneous or bizarre reactions take place in one corner of it, the effects in some miniscule way, at least, are felt throughout.

Surely this is, in itself, an esthetically satisfying picture of the Universe. The question, of course, is... is it true?

I think there is evidence it may be. Evidence which is increasing year by year, and to my mind, seems to be heading towards that Omega point of recognition.

So that I will not be labelled immediately as being out in some left field of the known Universe, let me give some indications of that direction in the remainder of this article.

First of all, Einstein himself never disowned the Ether. Quite the contrary...

"We may say that according to the general theory of relativity space is endowed with physical qualities," explained Einstein in 1920. "In this sense, therefore, there exists an ether."

Quite an admission.

Neither, for that matter, did Michelson and Morley ever disown the Ether. They did their experiment, left it for others to draw conclusions from it.

But let me give an example of a modern physicist who, I believe, though he himself may disagree with me, champions the cause of the Ether, or perhaps something like my concept of the Universal Medium.

I speak of T.D. Lee, Professor of Physics at Columbia University, and co-winner of the Nobel

Prize in Physics in 1957. In his small but very important book published in 1988, Symmetries, Asymmetries, and the World of Particles, he comes dangerously close to coming right out and declaring it.

I say dangerously close because, the world of modern physics being as it is, no different in some aspects from any other organization seeking its own perpetuation and self-promotion, and therefore subject to pressures and subtle intimidations which might as well carry a label, Don't rock the boat, Professor Lee will stop just short of absolutely declaring such a heresy.

Let me quote from pages 21 and 22 in a section titled "Vacuum as a Physical Medium"...

"What is a vacuum? We all know, for example, that the earth has an atmosphere. If we pump out all the air and all the matter, then what remains is the vacuum. But, insofar as we are incapable of switching off physical interactions, the vacuum could possess enormous complexity. As we shall see, virtual creation and annihilation of particle-antiparticle pairs can occur continuously in the vacuum state. Therefore, the vacuum resembles a physical medium.

"In the last century, in order to understand how the electromagnetic force, and later the electromagnetic wave, could be transmitted in space, the vacuum was viewed as a medium called aether."

Professor Lee then quotes Michael Faraday, who believed that magnetic force was indeed transmitted through and by the Ether, an Ether he conceived of as having other uses besides the simple transmission of forces.

Is that not interesting? Notice that Professor Lee all but identifies... I would say he <u>does</u> identify... what he calls vacuum with the Ether.

And he is clearly saying it is a <u>medium</u>, one that is capable of transmitting electromagnetic waves.

Of course it seems to me he is twisting very hard the meaning of vacuum. Twisting it so hard, it seems to lose its ordinary meaning. Because if the reactions he describes—virtual creation and annihilation of particle-antiparticle pairs—occur <u>only</u> in a literal vacuum, then that must mean they do not occur where there is air, which seems a mistaken concept. That would say such a reaction could not occur within a linear accelerator, where we know they do occur. The mere presence of air would never negate such a reaction from occurring. So why choose such a mistaken and misleading term? I would say fear of ostracism by his scientific cohorts for daring to advocate a concept apparently disproven and discredited many years ago.

But was it ever disproven, the Ether? I would say, no. It was merely discarded as unnecessary, or so considered at the time.

Professor Lee goes much further in his concept of the "vacuum".

He says on page 46,

"The two major puzzles that face us are (1) missing symmetries and (2) unseen quarks."

He talks of cases in the microscopic world where parity, a favorite and much coveted concept in an orderly universe, is apparently not maintained in a number of experimental tests.

What is the explanation of this violation of a fundamental scientific concept?

Professor Lee says the violation is perhaps only apparent. That if we take into account the "vacuum" then all apparent lack of parity and violation of symmetry can be explained. In other words, what takes place within this universal medium he calls the "vacuum" can complete all these questionable reactions, and in the process maintain parity and conserve all symmetries.

Let me quote his exact words, from pages 46 and 47 of the little section titled, "Two Puzzles".

"As we have said, symmetry implies conservation. Since our entire edifice of physical interactions is built on symmetry assumptions, there should be as a result a large number of conservation laws. The only trouble is that almost all of these conservation laws have been violated experimentally. This is the essence of the first puzzle, missing symmetry, which has been discussed before. As I mentioned in the preceding chapter, this difficulty could be resolved by introducing a new element, the vacuum. Instead of saying that the symmetry of all matter is being violated, we suggest that all conservation laws must take both matter and vacuum into account. If we include matter together with vacuum, then an overall symmetry could be restored."

He then goes on to explain the mystery of the missing quarks in the same manner. It is all due to the "vacuum" which somehow swallows up the quarks, keeps pressure on them and will not let them emerge as single entities.

And now back to the basic question raised earlier in this article.

What can unify the Macroscopic and Microscopic worlds?

Does it not seem obvious there must be a connecting medium?

Where a free exchange is made between the two worlds that otherwise indeed would be separate entities?

Einstein's dream of a unified field theory is nearing realization with the present progress of theoretical and experimental physics.

On the final page of his book, Professor Lee has this, to me, rather startling statement to make.

"To resolve these two major puzzles, missing symmetries and unseen quarks, we invoke the dynamics of the vacuum. If the vacuum is indeed the underlying cause for these strange phenomena in the microscopic world of particle physics, it must also have been actively responsive to the macroscopic distribution of matter and energy in the universe. Because the vacuum is everywhere and forever, these two, the micro and the macro, have to be linked together; neither can be considered a separate entity."

That is a very strong, even shocking statement in the sense of the revolutionary content behind the thought.

He is talking about nothing less than a missing link. The unifying factor between the micro-world and the macro-world.

The only reservation I have is with his terminology. What he calls the "vacuum", I would without reservation or shame call the Ether. It is, I think, an altogether more descriptive and more accurate term. A vacuum can not be all-pervasive, as Professor Lee describes it, but the Ether can.

We need to once again open our minds to it, the concept of the Ether. It held nineteenth century science together... it may well come to fulfill the same function for twenty-first century science. The New Ether may not be the same as its original counterpart, no. It could very well be both an absolute rest frame in the larger sense and at the same time an elastic, compressible and expandable and therefore locally moveable and malleable frame. If it combined these two qualities it might very well be able to account for the apparent invariability of the velocity of light, as well as for time dilation and other phenomena implied by Einstein's theories. Gravity itself might conceivably turn out then to be a property of the Ether. If massive bodies tend to concentrate the Ether within themselves, as do velocities approaching light speed, then much is simply and logically explained. And its all-pervasive quality per se unites the macroscopic and the microscopic worlds into a single Universe operating under the same symmetrical laws.

Even the indeterminacy principle might prove to be nothing more complex than the tendency of the Ether to smear things out in space as a part of its continual auditing and processing role in the Universe.

Philosophically speaking, too, the Ether could be satisfying. Either what we have instead of God, or perhaps, as Stephen Hawking likes to say, a clue to the essence of the mind of God.

In any case the idea itself may, in the context of scientific history, be something like Halley's Comet was to astronomy.

The Comet had appeared for centuries before 1682 when Halley gave it his name and an explanation of its nature along with a calculated orbit.

The great Comet skipped the nineteenth century completely but appeared twice in the twentieth, each time a little better understood by astronomers.

The concept of the Ether may be something like that, with its own periodicity. An idea whose time has come...

Again.

—THE END—

* * * * * * * *

Wandering Thoughts

Was my article published?

No.

In spite of the fact I sent it to various publications that should, at least, have been interested in the content, even though they might legitimately disagree with the conclusions reached.

What does that mean?

Nothing much. Just that if you have no reputation in the field... if your direction is different from what is current... well, you know the rest.

No dice.

Dice...

Which recalls again the oft-quoted comment from Albert Einstein... that God does not play dice.

Well... not on a macroscopic level, Albert, no.

But on a microscopic one... the answer is not only yes, but that playing dice is an integral component of the entire structure and working of the Universe. There is nothing bad about that, I should note... just as Evolution allows for mutation, which is certainly based on chance occurrence... so do the workings of the submicroscopic Universe.

The Random Element in atomic and subatomic physics is and must always be a fundamental part...

Randall Barron

Just to start an explanation, or an approach towards an explanation...

A million or so electrons, say, have no purchase in the macroscopic world. But in the microscopic... the atomic or subatomic world... they do. While a dozen do not. The universe in its workings can not pay attention to what a mere dozen electrons do or do not do under most circumstances. But a million... something inconsequential in our everyday, macroscopic world... that must be paid attention to. That must be incorporated into the total picture of what is happening now. And so Quantum Mechanics. And so the idea of Quanta... of a significant quantity of elements... of energy or matter... to make a change, to be registered on the graph of subatomic action.

The Quantum Prohibition... which does not allow small reactions to enter into the large world picture... at the same time works on a Random principle. In that Random principle large world physics can be violated. That is, the Universal

Medium presents a bewildering array of possibilities at all times. It has its own laws. When it liberates energy or matter into the macrosystem it is unrestrained... whatever it is that is going to come out, to eventuate into real time and space... may or may not be what we would logically expect.

The reason for this may remain for the time obscure.

Some of us would prefer the micro and the macro systems to be severely matched with no variation permitted.

And yet, a lot of progress is caused, perhaps, precisely by those unexpected and unanticipated variants...

Mutations are important... if we want to draw an analogy with Evolution and apply it to the Quantum World.

The microscopic rises up into the macroscopic to cause some observable change in our history. A change which we can then either incorporate and act upon... or reject, if it is within our power to reject.

In a way, we are being given on the microscopic level... the power to vary, and the power to choose... if we wish to do so.

In a way, we are being given respect... and being treated rather well.

But...

In a way, the Quantum system is a very conservative one, with built-in safeguards against accidental error.

But Einstein was never comfortable with it, even though he himself played a key role in its initial development.

Einstein's psychological outlook had a lot to do with the directions his work took.

Therefore...

Before I move on to more of my explanations about Doctor Einstein's Universe... which I have dared to call AAA Einstein... I want to discuss a few pertinent things. Among those things, are a phenomenon I like to call...

The Mysterious Doctor Einstein

Mysterious? You may ask...

With a big "but"...

BUT there was nothing mysterious about the life of Albert Einstein. It was an open book.

Wasn't it?

Well, yes and no. Mostly yes, however, concerning his personal life, though there were exceptions. But I speak here of his intellectual and scientific life. His life of the mind...

And as to that, I maintain there were mysteries... deep mysteries as yet unresolved.

I would like to take up a couple of them at this point because they are very germane to the thesis of this book, as you will see later on.

First is... $E=MC^2$.

This is the equation that every schoolchild knows today, but that no one in the world had even the ghost of an idea about before young Albert Einstein dared to set it down in print.

In that half inch of print he had the blueprint to change the thinking of the world.

I personally like to think of it also in a supplementary and different way. As the greatest poem ever written. Shakespeare, I am your greatest fan, and I apologize... but yes, I believe it is the Greatest Poem Ever Written.

It is poetic, if you think about it. Intuitive to the nth degree. It came from... where? Outer space? From the Great Beyond? It seems to have been an intuitive leap unmatched in the entire history of the world.

That may begin to give you an idea of what I think of the mind and thinking processes of Albert Einstein. But yes, even so, I do believe he may still be subject to criticism and to constructive revision.

How can I begin? Let me count the whys...

First of all, if you have what is probably the greatest single revelation ever given to a single human being in the history of the troubled and tortured world... why would you not give a complete exegesis and explanation to that world as to exactly how it came about?

To my knowledge and investigative capacities, Einstein never did.

Did some angel visit him in the night and show him these three characters engraved on a stone tablet? Or was he walking in the countryside when a nearby bush suddenly burst into flame and a voice cried out to him, in stentorian, authoritative tones, "Albert, the formula I am about to give you is of divine origin..."

Maybe something like that happened.

Maybe he had a vision and the magic formula suddenly appeared written across a stormy, cloud-tossed sky in electric letters...

Or in a dream he saw it... not only the formula itself but also the reality behind it.

Ah, yes, the reality behind it...

Given the efficacy of the formula itself, that is what is missing. Has always been missing...

The unasked question is... what did he see BEHIND the formula?

But I will ask that question here and now.

What DID Albert see?

Because he must have seen something. Something so convincing that it seared itself into his consciousness and would not let him rest until he had imposed it also across the resisting mind of humanity in general.

And what did Albert tell us, the ordinary citizens of the world about that colossal revelation?

It can be set down in a single cipher... 0.

That was the sum total of it.

Now what he told Nels Bohr or young Heisenberg or his wife... that is another matter. Perhaps he told someone on the Planet something about what was the trigger... what was the vision... what was the occasion... of the enormous revelation.

But if he did, that person or persons have not come forward to let us in on the secret.

Because... all social pretenses aside... this was a revelation so big... a mystic connection so special... that whole books could have been written about it alone.

Whole books SHOULD have been written about it alone.

But were not.

Because as far as I can tell Albert Einstein kept a tight-lipped silence about it.

He must have wondered, too... why don't they ask me about that? About how it came about... but they did not. They only wanted to know the bottom line. It is a formula... it works... and everyone falls in line with success.

Ours not to question why...

But that is exactly what I do want to question here.

Again, what did he see? This visionary scientist that was Albert Einstein?

In my own mind I think it must all go back to a vision of the Creation... to the origin of the Universe.

I have nothing to base this on but my own intuition.

What else can I use, since Einstein himself for his own unexplained reasons, never told us how or why he came up with his powerful formula?

Our Universe begins...

Everything propelled outwards... a cauldron of incredible temperatures... at those temperatures and pressures perhaps a free exchange between matter and energy.

A temporary condition, yes. One that assumed a direction as primordial matter settled out... as the temperature naturally fell from the cooling action of expansion... from the initial soup of mass and energy all propelled outwards from the point of origin of our Universe.

$E=MC^2$.

But... why squared? Why square the Speed of Light? What did that mean?

Because it had to mean something.

How did that exponent come to him and why?

Let me explore some possibilities. Very important ones for our own comprehension of the Universe.

Was it a reflection of the measure of distance travelled?

That the distance travelled by a unit of original mass transformed into energy in the Expansion that created Our Universe... in some unit of time... was equal to the Speed of Light times itself? Equal to the Speed of Light squared?

Is that what Einstein was getting at?

Was that the picture in his mind?

Was he saying that a unit of matter turned to energy in the furnace of Creation would be expelled outward to run along a course of C X C in length before its initial energy would be exhausted?

It might have well been something like that.

But why the mystery? Why not tell us? If that was it...

Or did he have a different picture in his mind?

Was he perhaps thinking of something like this... a unit of matter converted to energy as say, projected from a central source onto a flat surface? A flat surface which would be something like a movie screen, not rectangular but square... exactly C, 186,282 miles wide by exactly C, 186,282 miles high?

Maybe. I wondered about it. My own thoughts I composed into an article back in 1996, which I tried to present to the scientific world for its scrutiny.

Rather than try now to paraphrase it, I prefer to present the article itself. Here it is...

****

Randall Barron

December 20, 1996

An Einstein Proposal

—Randall Barron

We can probably all agree that Albert Einstein was one of the greatest scientists ever.

Does that mean, however, that all of his work was perfect?

Probably we can all agree that any scientist, however great, may be subject to being improved upon as more and more is discovered about the operation of the physical world.

With that established, I would like to take up some ramifications of Einstein's famous equation, the one that so shook the world, $E=MC^2$.

First of all, let us ask a basic question. What are the implications behind Einstein's formula?

The first and most important one, perhaps, is this. Energy and Mass are equivalent.

Second, the ratio between the two is astronomically large. That is, from a very small amount of matter, a great deal of energy can be produced.

Third. C, the speed of light, has some innate and very basic role in the relationship of energy to Mass.

From these three implications, other basic questions arise. The first, is the relationship of energy to mass as expressed in Einstein's equation supposed to be a mathematically exact one? Has it been accurately measured and proved to be so?

Why C2? What is the significance of the squaring of C? Are we being told that the ratio of the equivalence of mass to energy can best be visualized as an area? An area in space represented by an enormous square, each of whose four sides=C? That that area times the mass of the matter concerned, represents an exact mathematical quantity of the energy released when matter is completely converted into energy?

If so, there may be reason here to stop and think.

Because...

An enormous square in the sky is an extremely incongruous and awkward figure to represent such an advanced and inspired concept. I feel that way about it, even though the square of the distance plays a very prominent part of Newton's calculations of the force of gravity and in all the calculations of electromagnetic force fields.

A square is not what might be called a natural figure, but is, on the contrary. extremely artificial except when used to describe the diminishing of a force through distance.

So, do we dare suggest a substitution here? Of a more natural figure to represent this key ratio between energy and mass?

What if we just go ahead and do it, then examine the results afterwards with a critical eye.

191

The more natural figure that comes to me in the case of Einstein's famous formula is the surface area of a sphere. To begin with, it is a natural three dimensional figure as opposed to an artificial two dimensional figure.

The sphere is perhaps Nature's very most favored figure. We live on a sphere, among spheres.

Every point on the surface of a perfect sphere, of course, is equidistant from the sphere's center. Conversely, any symmetrical radiation proceeding from a central point, naturally tends to disperse in a spherical pattern.

For these reasons, and others, I have done a very daring thing. I have rewritten Einstein's famous magic formula, incorporating the surface area of a sphere into it to represent the ratio of energy to mass.

The formula for finding the surface area of a sphere with radius r is...

Area=4pi x r^2

(Area=four times pi times radius squared)

Now, if we let r=C, the radius equal the speed of light, we then have the surface area of an imaginary sphere being...

Area=4pi x C^2

(Area=four times pi times the Speed of Light squared)

And, putting this into Einstein's equation, we get...

E=4pi x MC^2

(Energy=four times pi times Mass times the Speed of Light squared)

While we have not gotten rid of the term C2, we have integrated it into the formula for the surface area of a sphere.

Randall Barron

The most obvious result of such a modified formula is we have increased the energy to mass ratio by exactly 4pi, or roughly 12.57 times.

Could this be correct? Has the exact energy conversion from matter ever been measured with exactitude?

If not, then I propose that it should be. The results will show whether this is indeed a closer approximation to the true conversion ratio of mass to energy, or whether Einstein is exactly correct, or perhaps the true ratio is something not yet correctly formulated.

My modified version of the world's single most famous equation does seem to me closer to the forms of nature. It brings geometry and time into the visual picture in a way the original does not.

Furthermore, the visualization seems easier. You may think of the conversion of a mass to energy in this geometric figure, with the energy radiating outwards from a central point. If we stop the clock after one second of time, then the outmost points of radiation will form the surface of a huge imaginary sphere of radius C, and this area multiplied by the original mass will give you the total amount of energy to be liberated by a total conversion of the mass to energy.

It is possible this may represent a more accurate figure than that arrived at by the simple squaring of the speed of light.

Note that this geometric image fits in very well with the imagery of the Origin of the Universe as now commonly accepted, the so-called Big Bang theory. Poetically and physically, it seems to me to have a chance at being correct. Of course, only exact and careful experimental measurement will say for sure.

—THE END—

* * * * * * *

Controversial Visions...

So there it is. My own thinking on the subject, which I went so far as to copyright shortly after I wrote it back in 1996. I also tried for publication in a scientific journal or two but... hardly to my surprise... was not accepted.

I am not so bold as to suggest my own version is more correct than Einstein's.

What has to be considered is... was his formula based on either of the two possibilities I have suggested? And I have to suggest them since he himself was silent on the subject.

If so, what it comes down to is this...

What happened at the origin of our Universe?

I believe firmly in my own version, as presented in some detail at the beginning of this book.

That Primordial Mass within one universe ended up creating a new universe... ours.

That Creation with all energy and matter radiating outwards from the Primordial Mass would, barring complications, end up producing a spherical universe.

Such a spherical pattern would also accord very well with the formula, or equation, I have presented in my quoted article.

So whatever the resolution is or may be, it seems to go back to the origin of the Universe.

At least that is the way that I see it.

But, to add more fuel to the flames...

As I have said before...

Where did Einstein's famous formula come from in the first place? Einstein in his writings expounds on many and sundry subjects, but he seems to have left $E=MC^2$ in the lurch, as far as giving us the inside dope on where it came from, how it developed in his own thinking.

* * * * * * *

Some time has passed since I wrote the immediately foregoing.

Sometimes in science, just as in any other controversial field, you have to be prepared to eat your own words. Which is why I keep salt and pepper and a plentiful supply of salsa always at hand.

In this case...

Now, after all my comments about the hidden mystic explanation, the visionary conception that Einstein must have experienced in coming up with his magic formula of $E=MC^2$...

I am brought considerably nearer Earth again by what I find in further investigations of my own.

Item...

Physics has had a Kinetic energy formula for a long time, which reads like this...

Kinetic energy = $1/2$ x M x V^2...

Doesn't that look hauntingly familiar?

V is velocity, of course, and M is mass.

MV^2 is almost the same as MC^2.

In fact, if velocity were the Speed of Light, it WOULD BE MC^2.

Hmmmmmmmm.

Okay. What next?

Just to say, well, it makes sense. To figure total POSSIBLE or POTENTIAL energy would you not take the maximum possible velocity? Yes, obviously. And if that maximum velocity were, which it is, the Speed of Light...

Then there you have it.

Total Possible Energy from any mass = Mass x C^2.

There except for the factor of 1/2 we have Einstein's famous formula.

Was it really all that simple? Was that how Einstein arrived at it?

If so, maybe we can understand why he didn't really want to talk about it much... it might have made it look too elemental, too easy, rather than as the flight of genius it seems when unattached to any precedent. Even an Einstein can have an ego.

Of course I haven't explained the 1/2 factor yet. That is maybe because I can't. Of course, Kinetic Energy as a concept is an attempt to define the energy produced by the motion of the Mass as compared to the rest mass of the object. For some reason, obscure to me, that in the formula figures to be Mass x V2 divided by 2. Why, I don't know. Obviously the energy of motion is not the entire potential energy contained in the mass, but why exactly one half? There has to be a simple explanation, but it doesn't yet come to me.

* * * * * * * *

I write this months later...

Now I see even a more direct and earlier explanation for how Albert Einstein came upon his monumental and earthshaking equation.

I almost shudder to present it. Why? Because it takes so much of the romance and mystery out of the concept...

To know that someone around 1700 had basically the same idea.

It is something like the effect of knowing that Edgar Allan Poe had the basic concept of the so terribly named Big Bang theory a hundred years or so before any modern scientist did...

Fascinating, but at the same time disillusioning and demythicizing too, in a way.

But here it is. For your examination and judgment. I would not be AAA Einstein if I failed to present it...

It is all so simple, really. But then, that is what I am about... the search for simplicity amid a sea of apparent complexity.

Nothing that your local attendant at the AAA Auto Club couldn't explain...

So here it is.

His name was Gottfried Wilhelm Leibniz. A name redolent with fame in many fields... philosophy, science, literature... politics. The co-inventor of Calculus along with Isaac Newton. Well, co-inventor no... they didn't work together... each produced a theory of Calculus independently, and during the same time period.

And somewhere around the year 1700 Herr Leibniz came up with this formula...

vis viva = M X V^2

Does that not look familiar?

Maybe not yet. But "vis viva" was a term to denote in Latin something like "living force". Which is very close to what later would be termed by science "energy".

And so then we have...

Energy = MV^2

Now it starts to look more familiar.

To make it Einsteinian, all you have to do, if you want to calculate the MAXIMUM ENERGY available in any given Mass... is to multiply it by the square of the MAXIMUM POSSIBLE VELOCITY in our Universe. That of course, is C^2... the Speed of Light squared.

And so... E = MC^2.

Seen that way, not really a new formula at all. A very old one, modified to express the maximum rather than just the actual in any particular case.

The only new element is the realization that the Speed of Light is the maximum achievable velocity in our Universe.

Seen that way, Einstein becomes less visionary, and the great Leibniz more so.

In other words, Leibniz, if he had known the Speed of Light and wished to calculate exactly what a maximum energy might be, could have written Einstein's famous equation... well over 200 years previous to Albert's monumental feat.

It makes you think.

Makes me contemplate the possible fact that my own concept of $E = 4pi \times MC^2$ might be yet more mystical and inspired than even that of the great Einstein. Based as it is on a spherical concept, and having to do more with the origin of the Universe.

Only then I have to figure out what happens to the excess energy...

****

Let's leave all that for the moment and move on.
What about this?

What if Einstein did derive his famous formula
as I have speculated here above as a variation or
substitution into the oldtime formula for Kinetic
Energy, or even better, from the earlier Leibniz'
equation?

And what if...

As a result, his formula shares any possible
defect contained in the original Kinetic Energy
formula?

Well...

What possible defects could I be talking about?
Hasn't kinetic energy been measured according to
this formula for centuries now? And hasn't
essential energy been measured according to
Einstein's formula ever since he devised it?
Notably in atomic explosions and in cyclotrons? And
doesn't it all square with the two equations as
they are?

The answer is yes. Apparently.

But appearances can sometimes be deceiving.

What I would say is this. What if the energies measured have been only directional or partial?

My own basis for measuring energy as reflected in my formula of $E = 4pi \times MC^2$ accounts for ALL energy.

All right, Randall. All right. Suppose what you say is true. What happened to the other unmeasured energy? Where is it? Why don't we measure it?

Good questions.

First of all, my concept for the correspondence of energy to mass is not directional. It is spherical. In all cases energy is radiated in all directions, conforming to a spherical distribution, not a limited directional one.

To advocate myself a little bit, this seems more natural, more corresponding to what we see and feel about Nature. We live on a spherical object in the Universe. We are governed by a sun which is spherical, and our fellow planets are all more or less spherical, too.

While we don't know for sure, the Universe may well be spherical, too. As I have elaborated here plainly, I obviously think it is. Raindrops and teardrops are basically spherical. The sphere is the most natural and most used shape in nature.

So it seems natural to me that any formula describing the projection of essential energy should by preference be spherical, too.

My formula is just the projection of the surface area of a sphere *one light second* into Creation... with the sphere being of a radius exactly equal to 186,242 miles. This seems to me a more natural description of what really happens when energy radiates outwards from any central point.

Okay. But how to justify this against the test of reality, and real measurements? Because I am describing over twelve times as much energy to be generated from a given Mass as what Einstein describes.

I mentioned apparent measurements. But what about unseen energy that remains unmeasured according to our present abilities to measure?

Ah, yes, what of that?

Now I need to call on my friend the Ether, or the Universal Medium, again.

If it really is the governor of the Universe... the regulator... the Stock Exchange...

Then does it not make sense that it can TAKE IN energy as well as disburse it?

And maybe that's what happens to all the excess energy released in any atomic explosion, in any cyclotron collision of atomic particles.

It goes back into the Universal Medium. It helps maintain the system and leaves energy stores for future disbursement in any part of our Universe... in the great and constant task of maintaining an equilibrium amidst the play of forces tending to destroy that equilibrium.

Does this sound too custodial? Does it smack of the G word?

Yes and yes, very deliberately.

Sooner or later we must all come face to face, even scientists, with an elemental truth. We are living in a Universe which shows every evidence of applied intelligence, of design.

Even of... heretic to science and redneck rebel that I am... programming.

Programming? Horrors!

Call the men in white coats! Quick!

Stop and think about it. We have a Universe that in its extension to us seems infinite, though it is not. It functions day to day and year to year and century to century. Millenium after millenium. It is here. It works. It survives. It endures.

All by accident?

By a pure conglomeration of chance occurrences?

Try to write a computer program of your own... any program... but careful because any mistake, any wrong step along the way, will make it fail. And as far as I know no one ever wrote a program for anything accidentally. Intelligence, knowledge, skill, intent and design were all involved, with still no guarantee the program would work or long endure.

If things customarily fell out so well accidentally none of us would have to work, or strive to advance ourselves through applied knowledge. We would just sit back and catch the fruit falling off trees.

In fact, sometimes I get the impression that scientists who see the evolution of humanity as based on a series of purely random accidents... and those who advocate the Big Bang theory and the evolution of our Universe as also based on a similar chain of accidental or random happenings... must live in some kind of scientific version of the Big Rock Candy Mountain. You know the song about a bum's Paradise... where the cops have wooden legs... the bulldogs all have rubber teeth... and the hens lay soft-boiled eggs.

Sure.

But things don't work that way. Not accidentally.

Only programming has purchase. The Universe is that way, too... at least according to my own theory.

The Governor or overseer of the program is the Ether or the Universal Medium. It governs the Macro World we know, and the Micro World we don't... and forms the link between Einstein and the Quantum scientists. It puts everything together, or will eventually, because it is the force which KEEPS everything together, keeps everything running.

It is interesting that the same Orthodox Science which lives in the Big Rock Candy Mountain, seeing everything in our present Universe as nothing more than a fortuitous combination of circumstances... at the same time comes up with nothing but the gloomiest of forecasts for the eventual Fate of that Universe.

Fascinating.

More later.

****

To all this I need to add a very important ingredient, which is this.

Early on, I took issue with Science for ever pinning a name like the Big Bang on something so important and portentous as the Creation of a Universe. I took pains to point out why it was not really an explosion at all...

In describing it in brief terms, yes, it SEEMS very like an explosion at first glance... but that is only at first glance. The reason for that is Time.

Its variability.

Even though the time of Creation may be described in some ways as taking place within a few instants, with the original components of the Primordial Mass being rocketed outwards at Light Speed, we have to always remember certain things.

One is that Time would not run in such a process as we know Time. An instant in a Black Hole becomes frozen... frozen into an eternity as measured by anyone outside the Black Hole.

And a Black Hole or something like it is what we sprang from. The Primordial Mass...

So Time at first in the Primordial Mass would not exist. As the Expansion began with portions of mass and energy being rocketed out radially at something like a considerable fraction of light speed... well, that may sound fast to us. But at the incredible densities still extant inside the expanding Primordial Mass, the Speed of Light may not have been very much at all. A second inside that Primordial Mass might very well be a thousand years as measured by us in our current Universe. With Time no longer frozen, no, but in a very slow and measured March, even though constantly growing faster...

Growing faster because as the Mass is projected outwards the density of the forming Universe constantly decreases. Each decrease in density of the Universal Medium makes Time run faster.

Which leads logically to an examination of the next question. Which is this...

How long did The Creation go on?

In terms of our own present time I propose this...

That we have no idea.

How could we? Time itself was being Created in the very process of the Creation of our Universe.

And that was a gradual process.

Again, how did it work?

My belief, the Primordial Mass was being radially projected in a serial manner.

That content consisted of Mass and enormous chunks of energy liberated from the nuclear conversion of Mass plus something else very important... something integrated and highly compressed into the Primordial Mass which the parent Universe itself contained.

What I am talking about here should no longer frighten or upset anyone. I am talking about a 4th Dimension. Not Time, which is really not a dimension at all, in spite of the popular and scientific obsession to consider it such. No, not Time. Time is not a dimension... it is a process.

I am talking here of a true 4th Dimension. But maybe, just to avoid confusion, since Time has come to be so widely regarded as a kind of 4th Dimension, I should always refer, as I have previously, to what I am talking about as being... the 5th Dimension.

The question naturally arises at this point, as to the relationship of the Universal Medium and the 5th Dimension. Are they both really the same thing? Or two things but fully integrated so as to be inseparable? Perhaps the Universal Medium is generated by the 5th Dimension?

These are important and cogent questions, but ones which can not at present be fully answered or understood.

An associated question is this...

Is the doorway to the 5th Dimension dimensionally restricted? It would seem to be so. That is, only subatomic particles below a certain size can find admission there first and later egress.

That would seem to make sense from many different angles.

As to what exactly that admission size might be poses an interesting question in itself. It would seem to be a quantity that might already have popped up in various ways in our Quantum calculations. Something perhaps like Planck's Constant, or some other mysterious number that haunts equations like some mischievous ghost who will not give away its identity but continues to tease by its presence.

And I am also aware that after admission into the 5th Dimension, it might conceivably be found there are other dimensions lurking there. That it is possible there could be an even smaller and more exclusive doorway or gate... But while aware of the possibility, I prefer to shave myself with Occam's Razor and look for no more complex explanation than is absolutely necessary to solve the problem.

Now...

What else can we say about the 5th Dimension?

The 5th Dimension comes from the Parent Universe, THE GREATER UNIVERSE, where it was probably very much a part of the overt workings of said universe.

Not so, with our offspring, upstart Universe. Where the said 5th dimension is so compressed as to be non-observable under ordinary circumstances. It is an underlayer of our Universe... of key importance to its working and our understanding of how it works... but so tiny and deep and far removed from any direct surface observation as to seem to us utterly remote and fantastic.

Yet with analogies we can approach an understanding of it.

Our TV sets work fantastically. Not from anything we can see. The basic physics is beneath our level of observation, though the TV certainly is not. Invisible electrons... is anyone in the modern world going to dare to say they are not there? Just because we can not see them? We certainly see the images they produce.

If my picture of what may have happened is clear, and is taken into account...

Then it gives us a different view of The Creation. And incidentally, I ask no pardon for the use of the term, The Creation. In using it I apply no specific religious implication. Yet I admit an implication may be there. But if so, I did not put it there. The Origin of the Universe itself applied the implication as an integral part of the concept. Unless of course you prefer to use instead of The Creation, another term...

... perhaps, The Accident?

Or The Mistake? The Blunder?

Those who thought of the term Big Bang in fact might be highly pleased with any one of those last three. And so they could doubly demean not only the act but ourselves as people who try to comprehend and write about it... and the rest of humanity, too. Of course there may be other more neutral terms... but in general it comes down to this...

Was the Act purposeful or purposeless? I would have to come down on the side of the first, while Orthodox Science would seem to favor the second. It is hard to see how they could be so purposeful in this intent to demean, living in their self-defined purposeless Universe.

But... it's a free country, built I would add by people who had no feeling they were moving through a purposeless Universe.

Now... that redneck broadside fired across the beams of Orthodox Science... let us move on and back more closely to the original subject.

Let's resume with a question already postulated. How long did The Creation go on?

Orthodox Science would have us believe it was a matter of a few instants. Even some tiny fraction of an instant.

I hope I have given anyone who reads this reason to believe it may have gone on much longer.

Someone somewhere in the future can even come up, perhaps, with an approximate estimate of how much longer...

But for now I can just say this...

The Primordial Mass, according to my vision of things, would continue to be propelled outwards... to propel itself really... from its position within the parent universe. What would be the limit of that propulsion, in terms of distance and Time? I speak now only of the Initial Stage.

Ah, I don't give myself any easy questions. This is unexplored territory, undiscovered country... where Angels fear to tread, here we have the spectacle of this fool being only too willing to rush in...

All right.

So be it.

I will.

Deep breath and...

Here goes...

What about this?

As long as the original propulsive force is extant, so continues the projection of the original Primordial Mass.

How long is that?

Well, as we know, Time is a variable, in fact is being constantly created along with the other physical qualities of our Universe.

But to make a guess... say a thousand years, which is a nice round number and easy to remember. But it could as easily be ten or ten thousand or ten million. Any such figure would only have relevance to someone timing the process from OUTSIDE Our Universe. *It took as long as it took*, and Time in that circumstance really was not relevant since that was the crucible that created Time...

The Creative Expansion continues during such time as the original Primordial Mass endures.

As it goes, not to be too repetitious I hope, the picture is this.

The creation of a steadily progressing sphere... which I believe to be the basic form of our Universe.

That is the Universe we live in. As far as form.

But when did it stop? The Creation?

Well... to be perhaps a little technical and arbitrary about it... and a little sensational, too... it hasn't stopped. This is why I earlier implied The Creation is not over. Not yet...

At the same time... back at the Ranch... in Another Part of the Universe...

****

I believe some remnant of the original Primordial Mass still exists. Holding down the center of things and keeping us from flying apart too rapidly. But not at last keeping us from eventually flying apart after all...

Because we are... now flying apart.

That is the meaning of the ACCELERATED expansion of our Universe.

As I have mentioned at the beginning of this book, we are evolving from what originally was something analogous to what we call a Black Hole...

INTO WHAT?

Into something that could then by analogy be called A WHITE HOLE.

All very simple, really.

Elementary, my Dear Watson...

The lessening density within our created Universe is now becoming less than the density of our parent universe, THE GREATER UNIVERSE that surrounds us.

And what is the consequence of that?

Also elementary...

And so our outer limits, our spherical rim, is being everywhere subjected to gravitational forces... attraction, pull... from the surrounding parent universe... which is now denser than ours.

****

More Controversial Ideas

Suppose...

That the Universal Medium itself needs to have nutrients fed back into it, too.

To be constantly recharged.

That it is an Energy and Matter Exchanger, yes. But part of that exchange is the constant feedback of excess energies. An automatic, built-in Regeneration process. One of those excess energies may be what is intrinsic in my own formula. That for whatever energy-mass interaction takes place in the universe, something like 12.57 times that is fed back into the Universal Medium... which keeps it functioning always as the Great Regulator it is.

Should this be true then both Einstein and this Redneck scientist could be correct. He in the ultimate observed reaction by any scientist in any laboratory here on earth, since said observation would not include the automatic feedback into the Universal Medium which might go undetected, at least as yet.

And myself perhaps correct, too, in the calculating of what is necessarily and automatically proportioned back into the working mechanism of the Universe... the Universal Medium.

So...

I include my own personal calculations here as a way station in the ongoing task of establishing scientific truth. And to show that I have at least spent some time in contemplating these very important basics.

****

One very important basic is Light itself... one we have to talk about constantly in our path towards understanding Einstein, and understanding the Universe.

One basic question might be described like this... is Light a wave, an electromagnetic wave traveling through a medium? Or is it a particle?

Or is it both?

Because we can also ask the question... what is water? Is it a solid or a liquid or a gas?

We all know the answer to that one. Water is a solid... and a liquid... and a gas. Which form it takes depends on what are the environmental conditions. Normally, a liquid... but cold enough, it becomes a solid... and hot enough it becomes a gas.

No surprise there.

Although I guess we could have science broken up into three opposing schools. One proposing water is a liquid... the second saying it is a solid... and the third claiming it is clearly a gas. But no, nothing like that could prosper in our modern and very cognizant society.

But what everybody knows about water does not necessarily translate to what everybody knows about light.

I would ask, if light can bounce back and forth as necessary and, depending on external conditions, be either a wave motion or a particle...

Is there any reason to be surprised about that?

It does not seem surprising to me.

But yes, light remains a central mystery in the riddle of the Universe. And while Einstein has taken us a long ways towards resolution, he has left behind him his own set of mysteries.

As I have hinted before...

And what no science book except this one will ever dare to mention to you is this...

Light itself is an outlaw.

What do I mean by that?

Just this...

While Light is a keystone in Einstein's theories, it has this unexplained anomaly about it...

It does not obey the rules postulated by Einstein's own theories!

Because if light has any Mass at all, then that Mass would become infinite at the speed light travels at.

But that can not possibly be true.

If it were, light would hammer at us... not only hammer at us, but in the process, obliterate us... the rays at the beach would kill us, not by possible eventual development of skin cancer, but by the heaviness of the direct hammer blow against our bodies.

Well, of course we all know that doesn't happen.

The question is, why not?

It is a basic postulate of the Einstein theories that ANY MASS must become infinite at the Speed of Light. Certainly Light itself travels at the Speed of Light.

Does it not?

So?...

To me it means Light can not be a particle. If it were it would hammer us all to death. I would not be writing this book... no one would be there to read it, either.

And no, I haven't forgotten anything I said previously about Light assuming both forms, wave and particle.

Yes, Light may and does *act like a particle* under special circumstances. But even as it acts like a particle, it isn't.

That is my solution to the riddle.

What it becomes for a brief period of time, instead of a smooth wave progression of energy, is a procession of energy packets or quanta. Which can act like particles briefly, but quickly return to their natural wave pattern when conditions change.

The same thing happens within the atom. An electron which is normally a wave function, a kind of energy cloud dispersed over the full sphere of its orbit... can be converted momentarily to something resembling a particle... a concrete and more spacially focused bundle of energy... if interfered with by outside forces, such as the inquiring Doctor Cyclops eye of an electron microscope. This keys a conversion from one state to another, much as a drop in temperature might key a conversion of water to ice...

We are familiar with the water-ice conversion and not shocked by it. But not with the Light wave-particle conversion and so we tend to see it as very strange...

Which I contend it is not.

I also contend that the normal state of Light is as a wave... an electromagnetic wave which needs a medium to travel through.

That medium is the much despised and scientifically totally rejected... but still valid... Ether, or Universal Medium.

* * * * * * *

At one point I felt so discouraged about the prospects of the scientific community ever accepting again the discredited idea of the Ether, that I wrote a story about it.

About what might happen to anyone who might seriously, from within the Establishment, propose such a theory.

Here it is...

****

Apostate

A *few stars broke through the cold and overcast sky, twinkled frigidly.*

"There's no ice on the river at least, thank God."

"Let's do it fast. And for God's sake, remember this night never happened. You were at home, I went

to a movie. From now on that is truth, the rest lies."

The two men bent to their dark work. By the dim light of the car trunk they took opposite ends of the inert body, lifted it out.

One of them gasped in surprise, "He was thin in life. Ectomorphic. How could he weigh so much dead?"

"My guess is the lead plates I put under his belt definitely help. We don't want his body to surface anytime this century or the next. And don't talk so much. God knows who might be listening."

"At two a.m. on a cold February night?"

"You never know."

"He could have been a good scientist, you know... Yendrek."

The second man looked around apprehensively, whispered his reply. "No. Never. He was too stubborn. And his ideas... crazy. Yendrek made his fatal mistake when he wanted to discredit the *university. And where would that leave us? Just ask yourself that."*

In silence now the two continued their macabre task, half feeling their way towards the dim massive presence ahead.

Soon they were at the river bank.

A silent, steady, then rising rhythm united the two men in swinging the slender, inert body between them.

One. Two. Three.

A splash, then nothing.

Nothing but the muffled ongoing flow of the river.

The distant, frigid visual stammer of a dozen stars was all the two men saw to break the monotony of the dark.

They were all right, home free, one of the two men thought as he glanced at the stars. If those stars, as it seemed, were their only witnesses. A burst of curiosity came at him. "What did you use to knock him out with? Chloroform?"

"Ether."

The other man tried to stifle the ironic laughter that overwhelmed him, was not entirely successful.

* * * * * * * *

Yendrek had come to the university like many others. To realize a dream. In his case, the dream of becoming a scientist. Of making discoveries, discoveries that would rock the scientific world.

He felt them. They were out there, all around them. Ready to be captured like apples falling off a Newtonian tree.

Only no one talked of Newton much anymore. He was a man from another century, another world. A mechanistic, macroscopic world, so antiquated and out of favor it almost might as well be flat.

Yet Yendrek admired Isaac Newton, would have liked to have worked with him. To have been part of the gigantic effort to simplify the mysteries of *nature and put them in their Victorian place... a place subject to the laws of simple mathematics within a clockwork wind-up world, tamed at last, with humankind in charge, the Circusmaster if not the originator of the show.*

Modern science, Yendrek felt, had gone off into a chaotic world which at times seemed ruled by anarchists and drug addicts. Sure, that might all clear up one day. There might be an Omega Point where everything would turn around for the better and come together in Einstein's dream of a Unified Theory at last.

The trouble with Yendrek was something caught his eye from the past. From the very university he had chosen to study at.

One of the glories of that university was the work of Michelson and Morley.

Those two illustrious faculty members had performed their famous experiment and... everything had changed.

Everything was suddenly up for grabs.

The Ether, Newton's Ether, the Ether every nineteenth century scientist had depended on, was suddenly cancelled out.

Made into a non-entity. An embarrassment to the world of science.

Not that Einstein ever denied its existence. He did not. But neither did he use it in his own equations.

And that, Yendrek thought, was a mistake. A great and grave mistake that scientific thought still was reeling from.

He wanted to be a theoretical physicist himself, and what came to him one night when he tossed and turned under the burden of that deep desire, was this...

Yes, there was an Ether. And it was so important. He had worked out a preliminary theory, some equations of his own.

And was almost overcome by them.

Unless he was way off base, he saw the true importance of the Ether.

What it did was tie together Quantum Theory and Einstein. It was the Missing Link, the Common Denominator that made Einstein's Dream of a Unified Theory at last possible.

Not only that.

There was something deeper than that by far.

It had to do not just with physics and cosmology and the origin and purpose of the Universe, but with justice. The Ether was something special, something very like the all-communicating spirit of the Universe. Touched everything, penetrated everything, permeated everything. In a certain way, ultimately righted all wrongs, settled all accounts.

The Ether was nothing less than the Ultimate Arbitrator.

What physicists had instead of God. Or maybe better than that. God's Umpire, which implied... an interested Owner somewhere... guiding the fortunes of his struggling team from afar.

Oh, it wasn't exactly the classical concept of the Ether. No, his Ether, Bill Yendrek's Ether, was somewhat different. Elastic, contractile, flexible... almost, but not quite, weightless.

But it served as the necessary Arbiter of the Universe. Exchanging energy and ephemeral bits of matter all over the Universe at all times in ways we were only on the verge of beginning to understand.

Sometimes Yendrek thought of his Ether as the Great Accountant in the sky, other times as a mighty Stock Exchange in which everyone and everything was invested.

Then again, when his thoughts rose higher than baseball and economics, as a kind of judge who administered the fairest, most rigorous physical justice ever conceived of. Where not a photon nor a meson, much less a quark or electron was ever, ever lost or left out of the Great Ongoing Equation.

The Great Equation that ruled over everything, and whose workings were carried out through the all-pervasive Ether.

That it was, in its essence, a religious as well as a scientific concept, Yendrek grew gradually aware.

Which only made him feel closer to Einstein. Closer to Newton, too.

One thing for sure.

He could not let his concept go, could not forget it, could not be talked out of it.

The university professors in the Physics department tried.

They used ridicule and the pressure of low grades.

Yendrek would not be moved.

One night they called on him en masse in his little studio apartment. Using the carrot and the stick, they alternately courted and threatened him.

But... Bill Yendrek would not be moved.

He threw things back at them.

Lots of things, hard things, things that hurt.

One was that the Ether might also well account for the so-called "missing matter" of the universe, as well as being the conduit for gravity.

But the thing that hurt the most was what he said about the Michelson-Morley experiment.

Anathema. Heresy.

He tried to tell them the Great Experiment had proved nothing. That the experimenters themselves were deluded, full of false scientific principles which simply were not valid. Not then, and not now. Not ever.

"Look at their premises," he said. "The detector in the forward path of the earth through space was supposed to somehow become in the course of the experiment farther away from the light source than the one at ninety degrees out to the side. They said it was because the forward detector would have moved on ahead due to the earth's motion while the light ray heading towards it would not."

Yendrek paused for emphasis, hoping to pick up some empathy from the professors in his small studio apartment. He did not, but at least he had their attention.

"But this is insanity. Under that reasoning, let me give you a scenario. A balloon hovers above the earth on a day with no wind.

According to Michelson-Morley, the earth will turn beneath it without the balloon itself being involved in the general turning. The balloon, you see, is like their light ray. So, in twenty-four hours the entire surface of the earth will turn beneath the balloon, a complete revolution of the earth at over 1,000 miles per hour. Gentlemen, we know this does not happen. Gravity prevents it, carries the balloon along with the rotating earth, with the earth being carried through space also in its orbital path around the sun because otherwise we would all slide off the earth and into space ourselves..."

They were looking at him in shocked silence. He saw the resentful expressions on every face.

"Gentlemen, I submit to you this experiment, this supposedly great experiment, was a farce and a fraud due to faulty thinking. That it proved nothing. And most of all, it disproved nothing. Certainly not the Ether."

They came at him hard and fast. Which made him think of another analogy.

He pitched it at them.

"Michelson-Morley are telling us that a pitcher throwing in the direction of the earth's motion will have a longer path to throw because home plate will have moved forward relative to the flying ball during the pitch. Does anyone here believe that?"

This time they stared at him in silence.

"And don't think that's the only false premise of this experiment. I have more, but this is the most important one, the one that throws it on the junkheap of scientific history."

****

The instant Yendrek lost the struggle against the two scientists, he knew he was being assassinated, knew he was doomed.

The next-to-last thing that occurred to him under the influence of the ether-soaked cloth pressed tightly against his face was something like a vision.

In a single second he seemed to see every jot and tittle of his precious research papers. The papers programmed automatically to be released onto the Internet. Not only did they contain all his data, his reasoning, his theories, his equations, but also the request. The request that if anything caused his premature demise or disappearance, that his list of likely suspects be reviewed, with all his stipulations as to who and why.

And then... under the surface of the dark river's deep waters...

The last thing that burned in Yendrek's brain before that brain ceased to function was an image.

An image very like a hologram.

Galileo. Galileo who moved his ancient lips and spoke.

"E pur si muove."

Yendrek didn't know Italian, but he understood. The Pope and his representatives might say the earth was the center of the universe and that it didn't move, that the sun and stars instead moved around it.

All right. But Galileo didn't have to go along with that. His signature might, but not the living magic of his lips, which could not help but move, too, in harmony with the truth inside him, even as the earth did around the sun.

In the dark waters a perceptive fish might have seen the slightest movement, the slightest ghost of a smile, what even might be called an ethereal smile on the pallid lips of Yendrek's lifeless body descending slowly...

... descending as if in the cradling arms of some great Universal Arbiter for Justice. Who would not allow even a single photon or meson ever to slip away unaccounted for.

And so would pass the torch through the invisible network once again into hands that eventually would cast its light over the entire world.

THE END

* * * * * * * *

Second Thoughts...

Is that a little extreme? My story?

I think so.

But it reflects how I felt at the moment.

As a story it is not much, except for that expression of a frustrated feeling.

A personal reaction.

Hamlet could say, "The Time is out of joint."

What can you say to indicate that a previous concept that has been discredited never should have been? That it is still valid...

Science suffers from a common delusion of modern society. That the direction of progress is always forward... always upwards and onward.

My dissenting voice says this... it is not necessarily and not always so... in this case it is not.

The Ether exists, or rather the Universal Medium... whether the creeks rise or not, whether school keeps or not... and no matter whether the scientific establishment acknowledges said existence or not.

Just as the sun is the center of our Solar system.

No matter how many times Authority said it was not. No matter how many times the flat and stationary Earth appeared to Authority to be its center...

****

We tend to think that Science knows. That what Science says officially is true. But...

We have to be careful here. Science in the past has told us many things that proved to be incorrect. That stones could not fall from space. That humans could never fly. That to exceed the speed of sound would prove fatal. That humans could never fly to the moon.

Science would like us to think that the so-called Big Bang Theory concerning the origin of the Universe, and the idea of the Expanding Universe is (a) a modern idea and (b) an idea given to the world by the workings of Science and the Science establishment.

There is no reason to doubt that.

Is there?

Well, yes, there is.

Not only to doubt it but to show irrefutable printed proof that it was not like that at all.

What was it like?

Like this...

The basic idea of the origin of the Universe was first proposed, not by a scientist...

No. It was proposed by a literary man. A writer. One who the world, sadly, has never given any credit.

His forlorn hope was that the world would react differently once he got his ideas in print.

That the merit of his ideas would prevail... that the science establishment would recognize them... latch on to them. And in the process turn his own difficult life around...

His disappointment must have been extreme, and perhaps led to his early and tragic death...

In an article I wrote many years ago and tried to get published... only to find the same rejection that caused the sad fate of the original author of the theory... I delineate the truth of the matter, deny it as the scientific establishment may wish to do but... in the end, cannot.

Here it is...

****

Who Originated the "Big Bang" theory of the Origin of the Universe?

—Randall Barron

The title forms what looks to be a simple question with a simple answer.

In fact, if we were appearing on Universal Jeopardy, it would not be surprising for us to to come up with exactly the title of this article if we were given an answer such as "George Gamow in 1948".

That might win something for us on Jeopardy, but is it accurate?

Oh, sure, others might talk of how Professor Gamow enlarged on an idea by Lemaitre, the idea of the primeval atom. Still others of how Hubble's law and its proof of an expanding Universe was the key to validation of the theory. And we might mention that as early as 1912 the American astronomer Vesto M. Slipher had noticed the spectral lines of galaxies tended to be shifted towards the red end of the spectrum.

All true enough. But what if I were to name someone else who developed more or less the complete theory in the 1840's? Not only developed the theory but published it in great detail. And yet is never credited with having done so.

But how? How could such a thing be? Oh, you say, maybe we are talking about some long-lost manuscript that never saw the light of day. No. Not at all. You can go to your local library and find it in bold black print, as could anyone else in the past one hundred fifty years or so.

Furthermore, the author of that manuscript is one well known to American readers. In fact, to the world at large.

However, many would find him to be a most unlikely candidate for author of the Big Bang theory and of the now proven hypothesis of the Expanding Universe.

Would, in fact, find him to be an unlikely candidate as any kind of theoretical physicist or Cosmologist. And yet, I submit, he was both, and indeed was the originator of both theories if truth be told and if justice be done.

But to admit that seems somehow to go against the American grain.

However, most would readily admit he was inventive. Some even say he was the inventor of the art of the short story, though that is surely an exaggeration. That he was the inventor of the modern detective story, however, there can be no reasonable doubt.

The Gold Bug alone would be exhibit A, and if that left still some doubts, then The Murders in the Rue Morgue should clear them up.

Yes, I am talking about none other than Edgar Allan Poe.

Yes, I am entirely serious in proposing that he finally be recognized for something he never yet has been. That is, as a brilliant theoretical scientist.

Of course, I need to prove my case. And it will be a pleasure to do so, in the hope of Poe's receiving at last the recognition in science he has so long deserved and never yet received.

I have an explanation as to why Poe has been so neglected as a scientist, and it is simple enough. People tend to categorize and label other people. This is a natural tendency, an intelligent one. Because it simplifies our world, gives us a basis on which to conduct our lives. As part of a general attempt to be organized, it is generally a commendable tendency.

As to Poe's case, the fact his brilliant cosmology was and is published in a book of literature just about explains everything. Had it been published in a scientific journal, very likely things would have changed for Poe and for the scientific world.

But it is time to make the case.

First of all, I will try to summarize Poe's writing on the subject as briefly and cogently as I can as an introduction.

Poe theorized there had to be a primordial particle. That from this primordial particle came every atom in the Universe. He describes in great detail how a force irradiated these atoms from the primordial center in all directions to form an ever expanding Universe.

Before going ahead, I will just pause to ask the obvious question. Is that not a formulation of the Big Bang Theory including the expanding Universe?

So far ahead of its time and so clearly and extensively stated it is nothing less than astounding that there has been no official recognition of his accomplishment.

If we stumbled across $E=MC^2$ in an old Nathaniel Hawthorne story, would we be equally blind to it? Perhaps yes, we would. But whether so or no, there is no doubt about the peculiar case of snow-blindness produced in all of us by Poe's brilliant thesis.

Yes, of course, Poe stood on the shoulders of giants, too. Isaac Newton's in particular. But what a panorama was revealed to him, and how far ahead he saw...

Let me elucidate on that.

He brings in Newton's ideas on gravity early on in his thesis, and ties it very closely into his cosmological theory.

To begin with he makes a quick analogy between the law of gravitational attraction and the irradiation of energy from the primordial particle. The dispersion of light and energy and matter that varies with the square of the distance from the source is compared with the effects of gravity that vary inversely with the square of distance.

From this he infers things that Newton himself may have thought of, but to my knowledge never expressed.

It seems to me that implicit in Sir Isaac Newton's gravitational calculations with its inverse square rule sleeps the latent idea of the Universe's origin.

That is, there must have been tremendous initial force holding the primordial particle together, force which is diminished by one thing, and one thing only... distance.

And how could such a thing come about, the initial separation? To ask such a question is almost equivalent to the formulation of a theory. A theory so self-evident we could compare its genesis perhaps on an intellectual level with the theory of Continental Drift. Running time backwards, with the given evidence, it seems almost hard not to come up with it. And if you do come up with it, you have the Big Bang theory and the Expanding Universe.

Newton did not, however, to my knowledge, come up with any such theories. Though only he and God know what he might have conceptualized in the precious privacy of his own mind...

But Poe did.

Edgar Allan Poe, the mystic poet, cryptography expert, and inventor of the detective story.

Perhaps there is nothing to be surprised at here. Perhaps there is more unity than diversity if we look at him objectively.

In my own mind, I see Poe examining the Universe as nothing more—and nothing less—than a cosmological mystery. One to be solved. And he put his considerable abilities to work to try to do just that.

I believe he succeeded admirably. The only mystery now is why he has not received the recognition he deserves for his accomplishment.

That accomplishment is spelled out in some 170 pages in <u>Eureka</u>, to be found in any edition of the complete works of Poe.

I will just add here a few things of possible importance. Poe's theory was complete. He did not leave it in any sense in a halfway state. He saw the Universe as expanding until it reached a certain point, a point at which the primordial impetus was no longer enough to overcome the counteracting forces of gravity.

At that point, the reverse process of the original creation would begin to take place. The Universe would begin to contract and over a period of eons of time would eventually return to the initial unity from which it began.

He did not see it ending there, however. Like some modern theorists, he saw the entire process being re-initiated by the very forces compressing everything once again into the primordial particle. A rebound effect would start everything anew, and the cyclical Universe thus might endure forever. In his time, without the benefit of modern astronomy, he thought perhaps the contraction phase might already have begun...

He also thought the original explosion or impetus of the primordial particle might well have been manifested in a series of graduated impulses rather than as a single burst. He saw these going from the greater to the lesser until the Universe was populated and defined by the energy and matter thus projected. He tied that in closely with Newton's laws of gravity, seeing the origin as being in principle nothing more nor less than the reverse of gravity, with the force proportional to the squares of the distances the matter and energy of each burst was projected.

While this seems to me a brilliant insight and would indeed explain on a conceptual level the basic reason why gravity seems to work the way it does, it also seems there was no real need for his having broken the original moments down into individual phases. That is, an explosion of any duration by its nature would eventually taper off and thus automatically obtain the results Poe talks about. And Newton's theory of gravity could still be perceived as immanent in the single irradiating explosion.

But that is the way Poe explains it. And I can not but be intellectually bowled over by it. Especially when you consider Einstein, the greatest scientist of modern times. That is, Einstein, in spite of all his genius and his almost supernatural insight into the nature of things, still ended up going against the implicit results of his own theories of relativity. Because the Expanding Universe was yet unknown at the time he formulated his theories, and because he therefore felt he must account for a Universe in a steady state, he added to his original formulas as an afterthought something he was to regret the rest of his life. What came to be called the Cosmological Constant, a kind of fudge factor which could hold the Universe together and keep it from expanding. And then when it was discovered the Universe was indeed expanding, Einstein in his own mind kicked himself all the way across its vast expanse for falling into the temptation and trap of feeling compelled to add to his beautiful theories such an unnecessary patchwork job. A patchwork job which cost him the honor of being the prophet of the Expanding Universe.

Perhaps Poe should receive that honor. It would take nothing away from the deservedly honored work of Gamow, Hubble, Slipher and Lemaitre.

Edgar Allan Poe, one of America's greatest and certainly most eccentric writers, fought most of his adult life against the forces of disease and death and poverty, always on the brink of some kind of annihilation.

For that reason it is all the more remarkable what he was able to do in the field of science. That in the Haunted Palace of his mind, among all the ghosts and ghouls and tortured souls, there was yet another secret room, one lit by brilliant light. Inside that room he saw clearly the origin and eventual demise of the Universe, as well as its instantaneous rebirth as part of an endless, eternal cycle, a theory held by a goodly number of modern Cosmologists. He also saw there may be any number of other universes, completely undetected by us.

Poe realized full well what he had done, and his own reaction comes out, first in the exuberant title of his work, and in a quote from Johannes Kepler which he includes in Eureka...

I care not whether my work be read now or by posterity. I can afford to wait a century for readers when God himself has waited six thousand years for an observer. I triumph. I have stolen the golden secret of the Egyptians.

Poe has waited more than a century now. Much more. Perhaps it is at last time for recognition.

God knows he received little of that during the agonized span of his short and tortured life, which ended in isolation, confusion and ignominy somewhere on the streets of Baltimore on October 7, 1849.

THE END

* * * * * * * *

The Universe Again...

It is time to move on and expound more on what is really the crux of this book... the simple explanation of the Universe... the AAA Einstein Universe already explicated briefly in the first pages of the book, plus my own subsequent additions and subtractions and changes. And I do believe that the explanation of the Universe and some of its basic laws can be surprisingly simple once we do a couple of things...

261

One is to... throw out entirely the idea of Curved Space.

I believe, as I have said clearly, this was a mistake on Einstein's part. A mistake that led him down a wrong road and made it very difficult for him to get his ideas across to the general public... or even to scientists whose specialty is Physics... in brief, to put it bluntly, to anyone.

The idea of Curved Space gets in the way of the Theory of Relativity rather than making it easier to understand. Furthermore, it corresponds to no proven reality, since all of Einstein's basic points can be made without its use if certain other ideas are allowed to take its place.

You could say Einstein threw us a curve.

And...

He struck us out with it. But that was not his objective at all... when what he really wanted to do was something not defensive but offensive. He wanted to hit a home run.

But, well, we can leave the figures of speech and get into the science. Yet this is not to deprecate the use of metaphor and simile in scientific work. Not at all. They can be of key importance, both in the formulation of theories and in their later explication.

I have already mentioned that $E=MC^2$ is perhaps the greatest poem ever written. An Epic poem, even though its extent is only three symbols and not more than nine words. And what is it in essence? A comparison... a kind of simile or metaphor since one thing is explained in terms of another, which is what happens in a simile or metaphor. Energy is like a bit of matter multiplied by how far light travels in a single second and then multiplied by that same distance again...

Or...

Energy IS matter. Matter IS energy. And here is the ratio...

Or...

Matter is frozen energy. And here is the ratio.

Or...

Energy can be locked up in matter. But when it is liberated... look out!... its unleashed anger is tremendous.

Or...

The Speed of Light is a key element in the origin and construction, and the workings of the Universe. I am not telling you why that is so, but it is.

Or...

Light speed multiplied by itself is the intermediary between Mass and energy. What does that mean? What does it mean beyond the obvious fact that the differential is great, far greater than anyone previously had ventured to think...

Or...

There are a lot more things that could be written, and perhaps you can think of some, which I would be glad to hear of by Email should you be so inclined.

Which is what, to me, makes $E=MC^2$ perhaps the greatest Epic poem ever written. Even though I dared to propose a revised version of that poem, which I have already presented to you.

But to get back now to the beginning point...

We have thrown out Curved Space, which I maintain was an error in Einstein's conception. It was what blinded his own great vision. It was what kept him from seeing that his insight into the correspondence between acceleration and gravity was (a) correct, but (b) never properly explained.

If Einstein could himself have grasped WHY the effects of acceleration and gravity were identical... we would not today be puzzled as to why they are.

It's a little like, to turn literary on you again, that verse by Omar Khayam...

Myself when young did eagerly frequent

Doctor and Sage, and heard great argument

About it and about... But evermore

Came out by the same door as in I went.

****

Einstein's Dilemma...

Einstein had a tremendous truth. He intuited it, he knew it. He had a Universal Tiger by the tail. But...

He could never explain it.

What was the popular song lyric..."the words got in the way"?

In this case... Curved Space got in the way.

It began when in his General Theory of Relativity he tried to stitch Gravity into empty space. The result, in my opinion, was both horrendous and misleading. He himself thought he had converted Gravity... previously a mysterious force... into just a part of the Geometry of the Universe.

To do that, even though his goal was simplification, he really had to complicate things.

He wanted us to imagine too much that did not seem worth imagining, nor worthy of his own otherwise high concepts.

As I said, he wanted to project Gravity onto empty space.

Wanted us to imagine that empty space shaped by Gravity. Empty space that had more curves and twists than a funhouse mirror before he was through.

With nothing to transmit this gravitational force. No medium.

Furthermore, the object of this force had no structure of its own. Empty space. Somehow it was supposed to force stars and planets into their orbits. Leave them but a singular orbit because if they tried to go outside that orbit they would be forced back. By what? By empty space which somehow had Gravity now stitched into it... into its emptiness.

And he brought in Riemannian geometry to support his point of view. Which was supposed to curve every cubic inch of space...

Einstein's resultant calculations of Gravity were not necessarily wrong. Basically, they seem correct. Just that by their very nature, they were so complex and difficult, so unwieldy and so far out of tune with any kind of reality that anyone could see or imagine, that...

We went to the moon and back with good old Isaac Newton, not Einstein.

That tells you something.

To look at the Universe through Einstein's lens of Curved Space with Gravity stitched into its non-existent fabric... that is something that unfortunately has not been at all productive.

It was and is a stumbling block, to use a Jewish reference here appropriate.

What would have gotten Einstein... and secondarily us... out of the dilemma, the *cul-de-*

sac he found himself in once he adopted the concept of Curved Space?

In my opinion, simply to have thrown out the concept, and substituted in its place... the idea of the Universal Medium.

So, I said to understand Einstein and make him AAA, we had to do two things.

One, throw out Curved Space.

Two, punch in to all equations the idea of the Universal Medium.

Now we have helped Einstein out tremendously. The equivalence of Acceleration and Gravity becomes immediately comprehensible, where before it was totally opaque.

Not only that, the connection between Relativity and Quantum Theory can be made. The Unified Field, Einstein's Dream, can be seen as attainable... in fact the Universal Medium not only makes it possible, it is in itself almost a Unified Field, or at least its medium, its container.

Randall Barron

The tortured and strained explanations of some parts of Einstein's theory can become superfluous and unnecessary. About different observers seeing different things and their measure of time... which become so bizarre as to be absurd.

All that can go to file 13, where it belongs.

Now, let me make it clear...

Under my plan, we do keep almost all of Einstein's theories and equations... except those specifically mentioned that may be thrown out.

Now...

Certainly we don't need his Riemann geometries of Curved Space. All those problems can be solved by plotting the varying densities of the Universal Medium. The curving of starlight can easily be plotted, even more easily seen by using the Universal Medium. The picture looks like this. The Universal Medium radiating outwards from the center of the sun forms a kind of second sphere of gravitational force which acts like a lens in bending starlight. Easily calculated, easily perceived, with no need for Curved Space nor Riemannian geometry.

And we certainly keep all of Einstein's Lorentz Transformations. Those were ingeniously conceived by the great Dutch physicist and mathematician, H.A. Lorentz, and masterfully employed by Einstein to complete his task of correlating everything to the Speed of Light... which he would never tell us, but I will, AS LOCALLY MEASURED.

Even though it involves mathematics, maybe we should look at the nuts and bolts of one of those Lorentz transformations, since they are of such basic importance.

I will have to write the formula out in words since my computer can't accept the complex array of mathematical symbols. Anyway, maybe that will make it easier.

We have talked about the increase of Mass with velocity. This is just a formula to show how that increase is governed by how big a fraction of the Speed of Light the velocity is. That's all there is to it, but it regulates everything according to C, the Speed of Light LOCALLY... and does so very well and very exactly.

The formula just says the rest Mass of the body should be divided by the square root of the number 1 minus the velocity of the body squared... over the Speed of Light squared. This result gives you the measure of the increase of Mass due to velocity.

Now, these formulas work very well and, to state it once again, relate everything... in this case the increase of Mass... to the speed of the Mass as compared to the Speed of Light. Mass and Time and the measure of length in the direction of travel all change proportionately as measured against the Speed of Light.

What happens if we make the speed of the Mass equal to the Speed of Light? Well, when you substitute that into the equation you get C2 over C2 which equals one. When you subtract that one from the other one, you get the square root of zero which is zero. You then have the rest mass over zero... which equals infinity...

In other words, if you could achieve the Speed of Light your mass would become infinite. And what happens then? Ah, for that we go back to my story and let Scott Diamond's strange experience speak to us. At least that is my own idea of a possible consequence...

But my point here is that Einstein's use of Professor Lorentz's equations are very useful and in no way are to be considered as invalid.

But I might ask here a deeper question...

WHY does the Mass increase with velocity?

????????

It would seem to me that answering that question should be of prime importance to Science, too... though I do not see it answered... or even asked... anywhere I have looked.

Well, yes, there is a lot of vague talk about increased energy. Even from Einstein. But that does not square with his own formula of the equivalence of energy and mass... $E=MC^2$.

Which leaves it as something of a puzzle. Because, after all, it should be clear that the speeding body adds NOT ONE SINGLE ATOM TO ITS MASS... so???

The simple answer lies in the Universal Medium. Remember that the UM carries the force field we call gravity and distributes it.

What happens with motion is that more Lines of Force become concentrated inside the moving Mass. This is a direct result of moving through a field of force. The higher the speed the more lines of force tend to be jammed inside the speeding body, which makes its effective Mass greater.

And here once more is my Master Stroke of AAA Einstein magic for reducing the complex workings of Our Universe to a level we can all understand easily.

Very simply, the main constituents of Our Universe... Mass, Gravity, Time, measurement...

function according to a single, simple governing principle...

THE DENSITY OF THE UNIVERSAL MEDIUM.

That relative density will determine Gravity, which in turn will determine and regulate effective Mass, and Time... and how we measure them.

The one constant measurement within any local system will always be the Speed of Light. This may sound like Voodoo Science, like Black Magic almost... but it is true and a very simple way to approach understanding of how the Universe really works.

And if, as I suppose, the Universe functions fundamentally in response to the presence of the Universal Medium, then...

We are being told something much deeper than anything yet said...

We are being told that Our Universe is not just an accident or result of a series of random happenings.

On the contrary...

Our Universe is planned and programmed and therefore certainly must have come about intentionally. Through design.

Because it is integrated and governed by uniform and omnipresent rules and laws.

Anyone who doubts this should consider the following...

There is no need to get into abstract metaphysical or religious reasoning here. No syllogisms of logic, no ragtag remnant sale of ancient and modern philosophy... no, nothing like that.

This is Triple A Einstein talking to you here, no one more elevated than that. If he can understand it, anyone can understand it.

I don't talk down to you. I talk *up* to you. Perhaps the only one in the world of science who does. And certainly the only one who considers it a privilege to do so. Unless you know different scientists than I do.

So...

I ask you this.

Just look at games. The revelation, the whole enchilada is included there.

Inside games.

Think about it.

You dribble a basketball... that is, you bounce it against the floor and... it bounces back. Not just once... but every time.

What does that tell you?

It tells me this...

Something is going on here. Something behind the scenes. That is... this world is governed by rules.

Just a short step further to conclude... I can interact with these rules.

I play, therefore I am... an active participant in the Universe.

That is the idea I want to not so surreptitiously introduce into your conscious mind now.

Your subconscious mind... that is a different question... the idea has been there all the time.

Mountains, lakes, grass, trees... a beautiful landscape... a gorgeous sunset... yes, those can produce under the right circumstances a kind of religious awe... a stumbling towards some kind of a realization about the Universe and who or what might be behind it.

Okay.

But...

For AAA Einstein... and perhaps for the majority of humans... where he may not be able to easily achieve the sensitive perception it takes to make such an artistic realization concerning sunsets... he can achieve something similar through participation in games.

And that's all you or anyone needs.

In basketball you try various shooting techniques. Some work, some don't. Some work better or easier or more consistently than others. The point is this. Universal feedback exists. It is real. You can't escape it.

Input A results in output A. Input B results in output B.

The Universe is there. Always there. To respond to however you choose to challenge it.

Who knows? Today or tomorrow or soon you may find the secret technique to insure you make 90 per cent of all the basketball shots you take. Alan Iverson is close... you may be closer. Because if you do discover it, you can use it... and reap all the benefits. That is what the Universe is saying to you and me and everyone...

It is a Program.

Interactive.

It interacts with you... with me... with everyone... impartially.

You develop a tennis forehand superior to all others... you can advance. The Universe will not stop you, but on the contrary help you. It is an impartial program.

One question here. The most basic and important of all. If this is true... and who can deny it?... then...

Does this not imply... as I have suggested previously... does this not imply the presence of... A Programmer?

I think so. I think all religious questions can be reduced to this. If not through an appreciation of beauty... some esthetic epiphany... then through the more basic epiphany offered everywhere by games.

Games are the easy touchstone between anyone who plays them and the Universe. It is as simple as that. *I play... therefore I understand.*

I understand that the Universe is not an accident. That the Universe is planned. More than planned. Programmed.

We all appreciate this subconsciously. Even the lowliest couch potato who religiously... the term is appropriate... does no more than follow the athletic contests on TV.

I suggest The Pope should try the games approach to understanding the Universe... he and all the other religious leaders... I suggest they could all learn something quite constructive.

Games and science share a common basis... something rooted in the very concept of the Universe. There should be no shame in admitting this... in acknowledging this relationship. But to date, only your **AAA** Einstein is bold enough to declare it.

The Universe IS programmed. Games and science prove it. We are programmed even before birth... DNA proves that.

Why should we be shy in admitting this simple basic truth? Why should we be shy to admit that we and the Universe are created?

We should not. If up to now some of us have been shy, well... now is the time to change.

And now on to something else...

Randall Barron

Einstein's inference that everything in the Universe relates to the Speed of Light is impressive enough in itself...

But when it is realized that beyond that statement, there is a deeper truth... that Light itself rides on what we can consider to be the Basic Texture of Our Universe... which is our by now familiar friend, the Universal Medium... then... then we do indeed see deeper.

We see that everything relates to that Basic Texture, that Universal Medium. It is what keeps Our Universe one Universe, united and functional, with every part in exact adjustment to the whole.

And that, Ladies and Gentlemen, is something that Einstein never told you... but that AAA Einstein will.

And has...

****

EPILOG

I promised in the beginning to make the main workings of Our Universe not only simple but...

1. Accessible

2. Atomic

3. Artistic

I hope I have accomplished that.

And now... the final word...

The End... is a Beginning

What will be the eventual Fate of Our Universe?

I have already talked about that up front...

But...

There is so much to say about it in addition to what I have already said. I will try to be brief... yes, but I must leave some room, too, for what I call vengeance and vindication.

First of all, I have a great desire for vengeance. I believe I have been harmed... that all of us have been harmed... by what I can only classify after long thought as one of the greatest crimes of Orthodox Science.

Orthodox Science in the 20th and 21st Century to date has turned into an oracle of Gloom and Doom as concerns Cosmology.

Can they deny that?

Listen to what they have to say to us. Over and over again...

That the Universe, Our Universe, is doomed. It must end in disaster, they say. The overall picture is this...

Our Universe will end in either (a) a deep freeze at absolute zero or (b) crushing incineration, a fiery furnace.

Not exactly a pleasing prospect, either one.

Not at all...

But, they say, that is on a scale of billions of years... it doesn't affect our thinking, does not affect us in any direct way.

Sure...

Why should our petty lives be affected by the Universe around us?

Or by the thoughts in the mind of whoever programmed it?

Okay...

If you say so...

But I believe all this does affect us, deep down. And deep down is exactly where it matters the most.

Especially when Orthodox Science also says other equally not exactly inspiring things.

That our Universe began in a fiery accidental explosion... which is, in my way of conceiving things, a way of making it a contemptuous event...

whose lack of significance and elegance is certainly reflected in the name... the Big Bang.

What we are left with is this... and only this...

That life itself evolved by a series of steps that were random and accidental, nothing more than a fortuitous combination of circumstance and chemistry.

What it all adds up to is this.

We are nothing.

Of no importance.

We count for nothing.

Another conclusion would be that if there is anything like a Creator, which they obviously doubt... then said Creator is treating us with the utmost contempt.

Our supposed consolation for all this is that... according to Orthodox Science... our terrible doom will happen many millions of years in the future.

That is supposed to make it all right.

But not for me. Because what is the truth hidden behind such supposed consolation?

That we live in an uncaring Universe. One that is out of control, careening towards certain disaster. We ourselves, only an unimportant accident, highly improbable and living on borrowed Time.

Do you like that?

Not such an encouraging prospect. Fatal to any religious impulses. Fatal to any ultimate uplifting thoughts. Fatal to any transcendental ideas of the conception and purpose of humankind.

There are other implications, all of them deeply disturbing.

Because if you really believe all that, it takes away your sense of purpose and leaves you listless, pensive and pessimistic.

That is the source for my desire for some kind of vengeance against Orthodox Science.

I accuse Orthodox Science of using the technique of propaganda known as The Big Lie. For at least the past seven or eight generations now. They just keep repeating the Doom and Gloom predictions for the Universe over and over.

You can find it in nearly all the Science books, referred to often in science fiction novels, reaffirmed at school and on your History Channel on TV, and mentioned in passing in the newspapers and by stand-up comics.

In short, coming at you in different forms from everywhere. What could be called Total Saturation.

With the result that...

Finally we all end up believing it... at least a little bit... even though we might not want to, even though we hate admitting it to ourselves. But what else can we do? When the expressed view of so many experts seems so unanimous that it would make us look like kooks or weirdos to voice disagreement.

And yes, this is a deeply ironic situation...

Because the two giants of Science were both deeply religious persons. Isaac Newton in the most conventional manner, Albert Einstein in his own uniquely personal way.

And of course I recognize that Science is not deliberately lying to us... scientists themselves simply do not yet realize that the future they project is not a true one.

But the noxious and deleterious effect remains...

So what is needed here is some action on the part of someone to set things right. Vindication.

And so...

This is where AAA Einstein comes to the rescue. Which is ironic in a sense, too... because he loves both science and technology.

But the necessary work must be done. So...

He dons his magic garb... a pair of freshly laundered coveralls with certain residual evidence of grease and oil spots... and with the deft employment of nothing more than a pair of pincer pliers pops Orthodox Science's balloon of Doom and

Gloom... and lets it fall in tattered rubberized remnants to the floor of the auto shop.

Yes. The thirst for revenge is quenched. And the taste of it is sweet.

It is no more than just retaliation for all the mental torture Orthodox Science has put us through for so many years.

Because...

Look again at the opposite picture I have given you.

Our beginning could be considered a Creative Expansion, not an explosion. Furthermore, this Creation was planned... and the workings of the Universe programmed.

And our ending?

Well, it will not be a fiery or a frozen Doom, but the opening of New Horizons. Not an ending at all, but a New Beginning.

A new vision in which we are no longer lost and alone in unfriendly space...

A vision that happens by the way to explain the otherwise inexplicable mystery of the acceleration of our expanding universe. Now everything is clear. As explained in the first few pages of this book, our outer galaxies are accelerating ever faster due to a very simple circumstance. They are being attracted by the gravitational force of THE GREATER UNIVERSE which surrounds us. Our own gravitational field has progressively weakened with Time as our Universe continues to expand... until at last its Gravity is becoming weaker than that of the parent universe. Eventually every galaxy of Our Universe will be incorporated into THE GREATER UNIVERSE.

What a difference from the picture painted for us by Orthodox Science wherein we must either freeze or fry!

In this vision, my vision, we are launched on a Great Journey... with no more of Science's sad stories or dirges to distract us from our goal.

With every reason to be able to anticipate the vindication of all the best dreams born of humanity's highest hopes.

And...

Just one more observation.

I have said as early as Part 1, the road to our Destiny is already in process, as the Creation... as Genesis... continues.

The accelerating expansion indicates it.

And interestingly it leads to something which by continued Biblical analogy we might call... Exodus.

As Exodus was in the Bible, to continue the analogy, so may this cosmological Exodus be for Our Universe. Towards a greater freedom, an enlarged scope of living.

We will eventually blend into THE GREATER UNIVERSE. Our Parent Universe. Maybe it will be something in the cosmological sense akin to... going Home.

There will be surprises.

Quite probably there the 5th Dimension, instead of being a submicroscopic, infinitesimally tiny and invisible rolled up cylinder, will be fully incorporated into THE GREATER UNIVERSE.

And who knows what that will mean?

No one, of course.

But it is entirely possible it may offer a vision of life of a greater amplitude, more liberal and more free... presenting vistas and opportunities greater than any we have ever dreamed.

In fact, if you could permit AAA Einstein to speak in a voice more refined than he ordinarily has... well, he might say something like this...

Arthur C. Clarke has popularized a metaphor that I would wish to employ here to make my own aspired-to meaning more clear...

Arthur has said that the emergence of life into space is something like the emergence of life from the earthly sea onto the land... a bold evolutionary step enormous in its importance.

An apt and brilliant observation.

What I have to say will be much more controversial and therefore open to question.

But it is this. In my own concepts, I see the Universe itself entering into a new evolutionary phase. That phase... blending into the Greater Universe which I believe surrounds us... may even come dangerously close to the concept of what earthly religions have called Heaven.

Where we will not see, as we do now, as through a glass, darkly... but on the contrary cleanly, completely and pristinely.

What I am trying to suggest here is this.

Beyond all Relativity... there may still at last be certain absolutes.

Furthermore, I believe that Albert Einstein himself believed this.

But he said in effect that since these absolutes are opaque to our limited observations... that we will ignore them.

But that was a mistake. In my opinion.

We can't... we shouldn't... ignore these absolutes.

They are the most important elements of the structure of Our Universe.

Behind and beyond Relativity are the even more profound absolute truths of the Universe.

Einstein himself felt this... as expressed in his oft-quoted, and previously referred to favorite statement..."God does not play dice." His absolute standard was the measured Speed of Light, to which all other phenomena had to calibrate and adjust themselves... and I have tried to explore in this book what may be the deeper reality beyond that.

I look at it like this...

The great Einstein's formulas and theories were a work in progress. He never regarded them as final. He did not live long enough to complete them.

He might have felt during those last long years of frustration, something like Omar Khayam himself did...

Randall Barron

Both of them at last, in my opinion, frustrated poets...

There was a door to which I found no key...

There was a veil past which I could not see...

I hope that through *AAA Einstein...* and *Beyond...* I may have opened another door... perhaps lifted a veil.

Who knows what the great scientist's unique insights into the workings of Our Universe under the right circumstances could have become...

A vision, perhaps, to light our way towards a glorious future. One we can still achieve.

—THE END—

ABOUT THE AUTHOR

Randall Barron has published various articles in the Shakespeare Oxford Newsletter and has addressed Shakespeare conferences in the United States and England. His Shakespeare Authorship Question website is now linked to the prestigious *Proquest Literature Online*. His three Shakespeare books are available through 1stbooks and the major Internet booksellers.

Another lifelong interest is Science, especially in the fields of Physics and Cosmology. Wanting to know the deepest secrets of Our Universe, he has over the years investigated many scientific theories. In this book he shares new and controversial ideas on Our Universe... its origin, how it works, what its future is. Which based on Einstein, but radically different, too. Original solutions to important scientific mysteries are presented.

Email the author at *Webrebel@Prodigy.Net*

www.ingramcontent.com/pod-product-compliance
Lightning Source LLC
Chambersburg PA
CBHW031822170526
45157CB00001B/153